HOW TO DESIGN, BUILD & USE
ELECTRONIC CONTROL SYSTEMS

Other TAB books by the author:

No. 1060   *303 Dynamic Electronic Circuits*
No. 1133   *The Active Filter Handbook*

Dedication

I am dedicating this book to my family, in the order that these persons entered into my life: my wife Marie, my son Peter, my daughter Mary, my son John, and my daughter Catherine.

Frank P. Tedeschi

HOW TO DESIGN, BUILD & USE
ELECTRONIC CONTROL SYSTEMS
BY FRANK P. TEDESCHI

TAB BOOKS Inc.
BLUE RIDGE SUMMIT, PA. 17214

FIRST EDITION

FOURTH PRINTING

Printed in the United States of America

Reproduction or publication of the content in any manner, without express permission of the publisher, is prohibited. No liability is assumed with respect to the use of the information herein.

Copyright © 1981 by TAB BOOKS Inc.

Library of Congress Cataloging in Publication Data

Tedeschi, Frank P.
    How to design, build & use electronic control systems.

    Bibliography: p.
    Includes index.
    1. Electronic control.   I. Title.
TK7881.2.T42   629.8'043   80-20683
ISBN 0-8306-9844-2
ISBN 0-8306-1229-7 (pbk.)

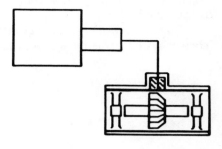

Preface

This book is for persons interested in understanding the theory, design, and application of control systems. At times the material becomes engulfed with mathematics; however, I assure you that the mathematics presented is the only way to illustrate the subject material.

The book begins with the classification of open-loop and closed-loop control systems, showing practical examples of each type. It becomes obvious that a special representation of control systems by block diagrams is necessary. Thus a chapter on block diagram algebra continues to simplify control systems within blocks that contain the components of the control systems.

It is necessary to illustrate the mathematics of control systems via Laplace transforms, which are presented in Chapters 3 and 4. Also, Appendices A and B are used to a great extent to aid the comprehension of Chapter 3. Chapter 4 presents the fundamentals of transfer functions employed to describe control systems. The emphasis in Chapter 4 is on Bode plots, and second-order and third-order transfer functions, which are quite commonly used in control systems. Chapter 5 makes use of the mathematics presented in the previous two chapters, when the stability problem of control systems is dealt with.

Control system components are introduced to the reader in Chapter 6. Only the most common components are dealt with in this chapter. A entire book could be written on the subject of control system components.

Chapters 7 and 8 describe practical control systems, such as position and velocity control systems using either DC or AC servo-motors. Also, the phased-lock loop control system, which is popular in communication and solid-state control system packages, is presented.

I hope you will have a better insight in the overall concept of control systems when you finish the book.

Frank P. Tedeschi

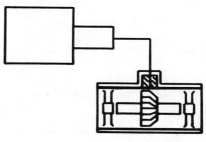

Contents

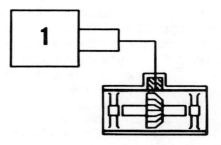

Fundamental Concepts

A control system is an arrangement of physical components connected or related in such a manner as to command, direct, or regulate itself and/or another system. You, the reader of this book, are a complex control system which contains five inputs (your five senses) and a variety of outputs, such as walking, talking, thinking, etc. Control systems regulate temperature in homes, schools, and all types of buildings. Control systems affect the production of goods, services, and food by insuring the purity and uniformity of the final product. The control system consisting of a man driving an automobile has components which are clearly both mechanical and biological. The major components of this control system are the driver's hands, eyes, and brain, and the automobile.

From the previous examples, it is clear that a great variety of components may be a part of a single control system, whether they are electrical, electronic, mechanical, hydraulic, pneumatic, human, or any combination of these. The desired result may be, for example, the direction of an automobile, the temperature of a room, the level of liquid in a tank, or the pressure in a tank. The regulation of energy is the key to achieving control. Control systems always involve changing conditions, and changing conditions always involve energy changes.

CLASSIFICATION OF CONTROL SYSTEMS

Control systems are classified into two general categories, *open loop* and *closed loop*. An open-loop control system is one in which the control action is *independent* of the output. A closed-loop control system is one in which the control action is *dependent* on the output.

An open-loop control system does *not* use a comparison of the actual result and the desired result to determine the control action. In other words, the action of the open-loop control system is determined by a calibration or setting to obtain a desired result. The same result will continue until the calibration or setting is changed from outside the system.

The firing of a rifle bullet is an example of an open-loop control system. The person holding the rifle aims the rifle to a desired target and fires the rifle. The actual result is the direction of the bullet after the gun has been fired. The open-loop control occurs when the rifle is aimed at the target and the trigger is pulled. Once the bullet leaves the barrel, its's on its own. If a disturbance, such as a gust of wind occurs, the direction of the bullet will change and no correction can be made to direct the bullet to the target. Figure 1-1 illustrates a block diagram of the rifle bullet being shot as an example of an open loop control system.

A toaster is an example of an open-loop control system because it is controlled by a timer. The time required to make the toast must be estimated by the user. Control over the quality of the toast is removed once the timer is set.

The primary advantage of open-loop control is that it is less expensive than closed-loop control systems. The disadvantage of open-loop control is that errors caused by unexpected disturbances are not corrected.

The closed-loop control system referes to the loop connecting the input with a portion of the output created by the feedback path. *Feedback* is the action of measuring the difference between the actual result and the desired result, and using that difference to drive the actual result toward the desired result. The term, feedback, comes from the direction in which the measured value signal travels in the block diagram, as shown in Fig. 1-2. The signal begins at the output of the controlled system and ends at the input to the controller. The output of the controller is the input to the controlled system. Thus the measured value signal is fed back from the output of the controlled system to the input.

The block diagram of the closed-loop control system in Fig. 1-2 will be employed throughout this book. The following four operations form the basis of the feedback control system.

☐ Measurement of the controlled variable.

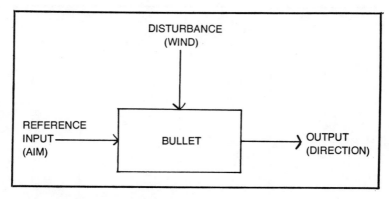

Fig. 1-1. Example of an open-loop control system.

☐ Computation of the difference between the measured value of the controlled variable and the desired value (the error).

☐ Use of the error to generate a control action.

☐ Use of the control action to drive the actual value of the controlled variable toward the desired value.

The process block represents everything performed in and by the equipment in which a variable is controlled. The process includes everything that affects the controlled variable except the automatic controller.

The feedback block (H) senses the value of the controlled variable and converts it into a usable signal. Although the feedback block is considered as one block, it usually consists of a primary sensing element and a transmitter. A typical feedback block could consist of a primary element, such as a thermocouple or a resistance element and a transmitter. The *thermocouple* converts a temperature into a millivolt signal, and the *transmitter* converts the millivolt signal into a usable electric current. A *resistance element* converts a temperature into a resistance value, and the *transmitter* converts the resistance value into a useable electric current signal.

The summing point or error detector computes the difference between the measured value of the controlled variable and the desired value (the reference input). The difference is called the error (E) and is computed according to the following equation.

$$E = R - B$$

That is, error is equal to the difference between the reference input and the feedback signal. The *reference input* or *setpoint* (R) is the

11

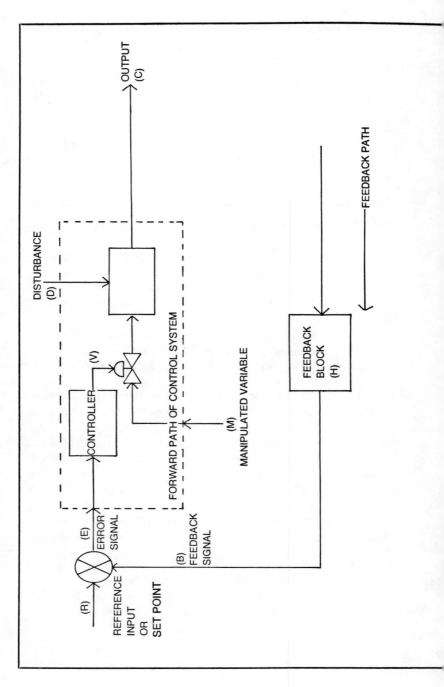

12

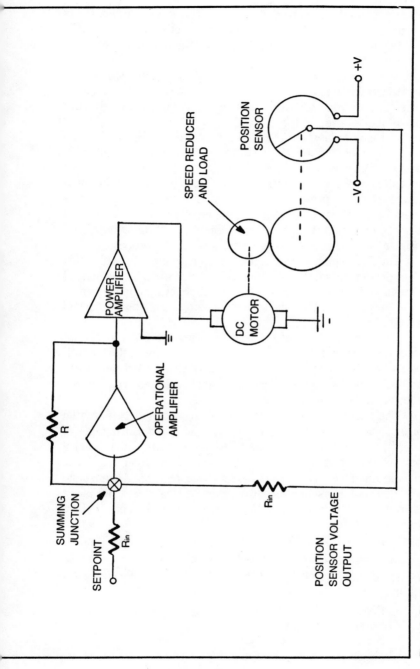

Fig. 1-2. Block diagram of a typical closed-loop control system.

desired value of the controlled variable. The *feedback signal* (B) is the measured value of the controlled variable or system output (C). The feedback signal usually differs from the actual value of the output by a small amount.

The *controller output* (V) is the control action intended to drive the measured value of the controlled variable toward the setpoint value. The *control action* depends on the error signal (E) and on the control modes used in the controller.

The *manipulated variable* (M) is the variable regulated by the final control element to achieve the desired value of the controlled variable. The manipulated variable must be capable of changing the controlled variable. The manipulated variable is one of the input variables of the process. Changes in the load on the process necessitate changes in the manipulated variable to maintain a balanced condition. For this reason, the value of the manipulated variable is used as a measure of the load on the process.

The *disturbance variables* (D) are process input variables that affect the controlled variable but are not controlled by the control system. Disturbance variables are capable of changing the load on the process and are the main reason for using a closed-loop control system.

The *controlled variable* (C) is the process output variable which is to be controlled. In a process control system, the controlled variable is usually an output variable which is a good measure of the quality of the product. Typical control variables are position, velocity, temperature, pressure, and flow rate.

The primary advantage of closed-loop control is the potential for more accurate control of a process. The two disadvantages of closed-loop control are a higher cost than an open-loop control system and the inherent problem of the closed-loop system to become unstable because of the feedback feature. An unstable system produces oscillation of the controlled variable which is not desired. However, the stability can be avoided. This will be studied later in the book.

TYPES OF CONTROL SYSTEMS

In this section we will classify control systems according to the type of application they perform. The four classifications we will use are listed below.

☐ Process control systems
☐ Servomechanisms
☐ Sequential control systems
☐ Numerical control systems

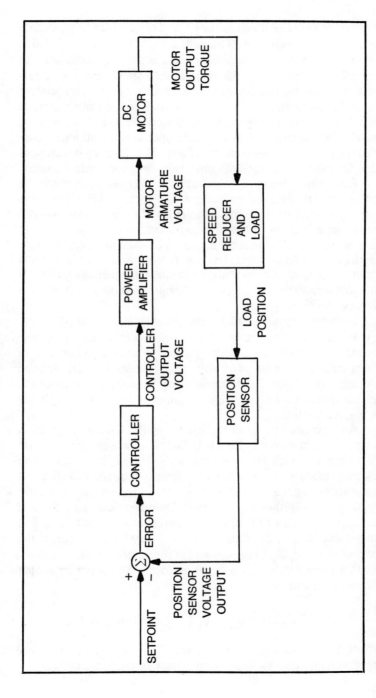

Fig. 1-3. A DC motor position control system. Schematic is at A, and block diagram is at B.

15

Process control involves the control of variables in a manufacturing process. Examples of process control systems are found in steel mills, automobile assembly plants, and food processing. The manufacturing processes change relatively slowly and are characterized by time constants of minutes or hours. The primary process control engineer usually does not have a precise mathematical definition or model of the process to work with, and the process model often changes with different operating conditions. Consequently, the process control industry has developed standard, very flexible controllers with one, two, or three control modes.

A *servomechanism* is a feedback control system in which the controlled variable is mechanical position or motion. Some process control systems are servomechanisms, while others may contain a servomechanism and a closed-loop process control system. The same mathematical elements are used to describe each system, and the same methods of analysis apply to each; however, because servo control and process control were developed independently of one another, each has a different design method and at times a different terminology.

Servomechanisms usually involve a relatively fast process; that is, the time constants may be considerably less than one second, compared to a process control system time constant of hours and even days. Usually, the system is well defined mathematically, and the controller may be designed to meet the system specifications by utilizing equipment that is most suitable for the application.

An example of a servomechanism is the DC motor position control system shown in Fig. 1-3. The position sensor is a 10 kΩ potentiometer with no stops and a 20-degree dead zone. The position sensor voltage output goes from a negative voltage ($-V$) to a positive voltage ($+V$) as the load rotates from its +170-degree position to its −170-degree position. The operational amplifier and the three resistors form the error detector and controller. The power amplifier is the final control element. The output of the controller is $-R_f/R_{in}$ times the sum of the setpoint voltage and the position sensor voltage. In equation form, the controller output voltage is as follows:

$$V_c = \frac{-R_f}{R_{in}} \left[V_R + V_B \right] \qquad \textbf{Equation 1-1}$$

where V_c = controller output voltage, V_R = setpoint or reference input voltage, and V_R = position sensor or feedback voltage.

The power amplifier increases the voltage and the power of the controller output. The motor is a permanent magnet field, armature-controlled, fractional-horsepower DC motor. The motor speed is proportional to the voltage applied to the aramature by the power amplifier. When the sum of the setpoint and position sensor voltages is zero, the controller output voltage is zero and the motor speed is zero. When the above sum is not zero, the controller output voltage is not zero and the motor speed is not zero. The motor will always rotate in the direction that will drive the above sum toward zero. Notice that the negative feedback in this example is accomplished by making the sign of the position sensor voltage opposite the sign of the setpoint voltage.

The DC motor control system in Fig. 1-4 uses the same controller, power amplifier, and motor as the DC motor position control system shown in Fig. 1-3. The position sensor is replaced by a speed sensor, which is normally a tachometer generator. A *tachometer* is a transducer that converts mechanical rotation into an electrical signal, such as a voltage. Thus, the tachometer produces a voltage proportional to its speed. The negative feedback is accomplished by connecting the speed sensor so that its voltage is opposite the sign of the setpoint voltage. In this control system, a small error signal is required to provide the armature voltage necessary to maintain the desired motor speed. The preceding examples are universal servomechanisms and will be discussed and analyzed in more detail throughout this book.

A *sequential control system* is one that perform a prescribed set of timed (sequential) operations. The automatic washing machine is an example of sequential control system. The control system of a washing machine first fills the tub. After the tub is filled, the clothes are washed. Then the tub is drained, and the clothes are rinsed. Finally, the washing machine spin dries the clothes. A few of the electromechanical components frequently employed in sequential control systems are shown in Fig. 1-5.

The opening and closing of electric circuits occur in all sequential control systems. The momentary contact pushbutton switches are shown in Fig. 1-5 as either normally open (NO) or normally closed (NC). A normally open switch will close the circuit path between two terminals when the pushbutton is depressed and will open the circuit when the pushbutton is released. A normally closed switch will open the circuit path when the button is pushed and will close the circuit path when the button is released.

The limit switches are actuated by a cam that operates the actuating arm. A normally open limit switch will close the circuit

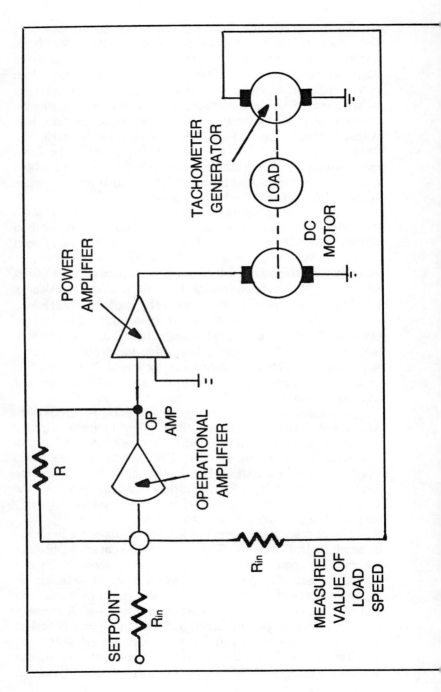

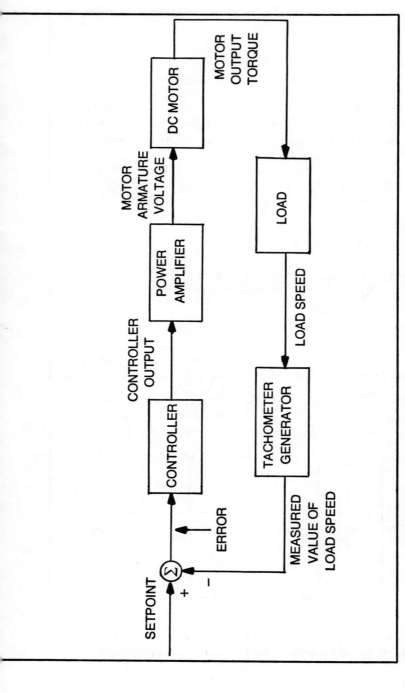

Fig. 1-4. A DC motor speed control system. Schematic is at A, and block diagram is at B.

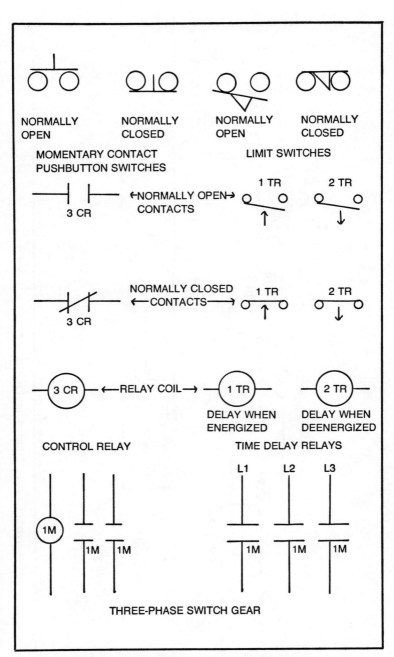

Fig. 1-5. Symbols of switches and relays used for sequential control.

path between the two terminals when it is actuated and will open the circuit path when it is deactuated. A normally closed limit switch will open the circuit path when it is actuated and will close the circuit path when it is actuated and will close the circuit path when it is deactuated.

A *control relay* is a set of switches that are actuated when electric current passes through an electromagnetic coil. The relay coil is represented schematically by the circle with the designation 3CR. The CR represents a control relay, and the 3 is a number designation to distinguish between two or more control relays. The relay is actuated or energized by completing the circuit to the relay coil. The switches are represented by parallel lines with the 3CR designation. The designation is necessary because the relay contacts may occur anywhere in the control system. The designation identifies which relay actuates a particular set of contacts.

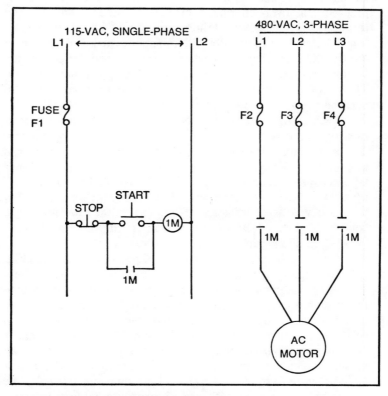

Fig. 1-6. A control circuit for starting a large AC motor.

Normally closed contacts are designated by a diagonal slash connecting the two contacts. A normally closed contact will open the circuit path when the relay coil is energized and will close the circuit path when the relay coil is deenergized.

Control relays that have provisions for a delayed switching action are called *time delay* relays. The delay in switching is usually adjustable, and it may take place when the coil is energized or when the coil is deenergized. The time, delay relay, is identified by a number followed by a TR. In a time delay, the delay occurs in the direction indicated by the arrows.

An example of a sequential control system is shown in Fig. 1-6, which is the schematic for an AC motor starting circuit. A motor starter for starting and stopping a 480V AC, three-phase motor will clearly illustrate a sequential control system. Pushbutton switches in the 115 VAC control circuit are used to energize and deenergize the relay coil (1M). When the start button is pressed, coil 1M is energized, closing all four 1M contacts. The three large contacts connect the three 480 VAC lines to the motor. The small contact in parallel with the start switch is used to hold the circuit closed after the start button is released. When the stop button is pressed, the circuit breaks and all four 1M contacts open. The circuit remains deenergized after the stop button is released because the 1M holding contact is open.

A *numerical control system* (N/C) uses predetermined instructions (program) to control a sequence of manufacturing operations. The instructions are usually coded in a symbolic program and stored on some type of storage medium, such as punched paper tape, magnetic tape, punched cards, or floppy disks. The instructions consist primarily of numerical information, such as position, direction, velocity, cutting speed, etc. Hence, the name numerical control. The manufacturing operations include, boring, drilling, grinding, milling, punching, routing, etc.

A simplified explanation of the N/C process is shown in Fig. 1-7. The engineering drawing completely defines the desired part. The part programmer uses the engineering drawing to determine the sequence of operation necessary to produce the part. He also specifies what tools will be used and how they will be used in relation to cutting speeds, feed rates, etc. The part programmer records his decisions in the form of a symbolic programming system used for this purpose. The computer processes the symbolic program to produce the control medium—for example, the tape used to control the machines. The computer performs many of

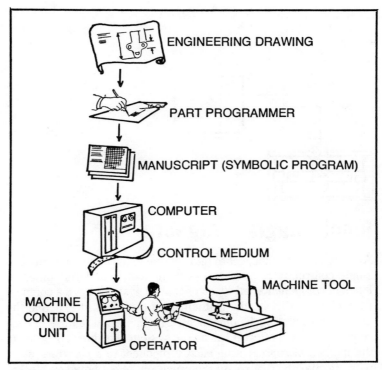

Fig. 1-7. A simplified explanation of the N/C process.

the calculations required to define each operation of the machine. The machine operator prepares the machine for operation, loads the control medium, and starts the operation. The numerical control system follows the instructions on the control medium to produce the finished part. The operator monitors the operation and checks for possible malfunctions.

A collection of terms and definition used in control systems is listed in the Glossary. You might want to refer to the Glossary before continuing with the book in order to familiarize yourself with the terms and definitions used in this book.

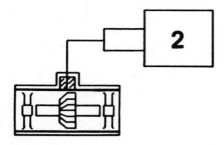

Block Diagram Algebra

Although it is not unusual to find several kinds of components in a single control system, or two systems with completely different kinds of components, all types of control systems have a common denominator; that is, the mathematical equations that define the characteristics of each component. The important feature of each component is the effect it has on the system. The block diagram is the method of representing each component in a control system with signal paths that relate the cause and effect between blocks, sets of blocks, and the input and output of a control system.

Each component receives an input signal from some part of the system and thereby produces an output signal for another part of the system. The signals can be electric current, voltage, air pressure, liquid flow rate, liquid pressure, temperature, speed, acceleration, position, direction, and any other physical occurence. The signal paths can be electric wires, pneumatic tubes, hydraulic lines, or anything that transfers one form of energy to another form of energy, such as accomplished in a transducer. Figure 2-1 illustrates a simple block diagram of a person driving a car.

The relationship between the input signal and the output signal is the most important characteristic of a component, and the overall characteristic of the control system itself. The relationship is called the *transfer function* of the component or the control system. In order to produce the output signal, a component may change the input signal in two ways, through size or timing.

A change in size is usually expressed as the change in the output signal corresponding to a unit change in the input signal, and

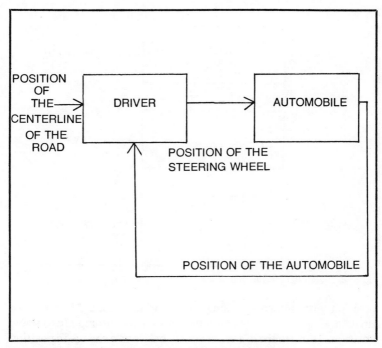

Fig. 2-1. Control system of a person driving an automobile.

is called the *gain* or *magnitude* of the component. If the input and output signals have the same units, a change in size means that the output signal is smaller or larger than the input signal. If the input and output signals have different units, the change in size must include the change in units.

A change in timing means that all or part of the signal is delayed so that the output signal occurs at a later time than the input signal. A component with a pure dead time characteristics will delay the entire signal by the same time period, which is called the *dead time*, t_d.

Figure 2-1 shows a simple block diagram of a person operating an automobile. The driver's sense of sight provides two input signals, the position of the automobile, and the position of the center of the road. The drive compares the two positions and determines the position of the steering wheel that will maintain the proper position of the automobile. To implement the decision, the driver's hands and arms move the steering wheel to the new position. The automobile responds to the change in steering wheel

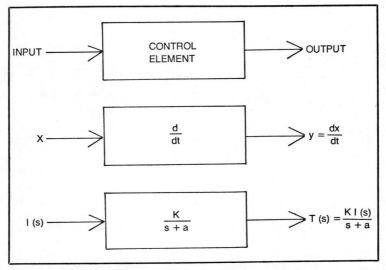

Fig. 2-2. Block representations of control system components.

position with a corresponding change in direction. After a short time has elapsed, the new direction moves the automobile to a new position. Thus, there is a timing delay between a change in position of the steering wheel and the position of the automobile. This timing delay is included in the transfer function of the block representing the automobile.

The loop in the block diagram indicates a fundamental concept of control. The actual position of the automobile is used to determine the correction necessary to maintain the desired position. This concept is called *feedback*. Control systems with feedback are called *closed loop control systems*. Control systems that do not have feedback are called *open loop control systems* because their block diagram does not have a control loop.

FUNDAMENTALS AND DEFINITIONS

The block diagram provides a convenient method for defining functional relationships among the various components of a control system. The interior of the block represented by a rectangle usually contains a description of or the name of the element, as shown in Fig. 2-2. The arrows represent the direction of unilateral information or signal flow.

The operations of addition and subtraction have a special symbol, as illuatrated in Fig. 2-3. The block diagram representation is a small circle with a cross in the middle. The circle and cross is called a summing point, with the appropriate plus or minus sign associated with the arrows entering in the circle from any direction.

In order to employ the same signal or variable as an input to more than one block or summing point, a take-off point is used as shown in Fig. 2-4. This permits the signal to proceed unaltered along several different paths to several destinations.

The blocks representing the various components of a control system are connected in a fashion which characterizes their functional relationship with the system. The basic configuration of a closed loop control system is shown in a block diagram in Fig. 2-5.

It is important that the terms used in the closed loop block diagram be clearly understood and remembered. Lower case letters are used to represent the input and output variables of each element as well as the symbols for the blocks, when these quan-

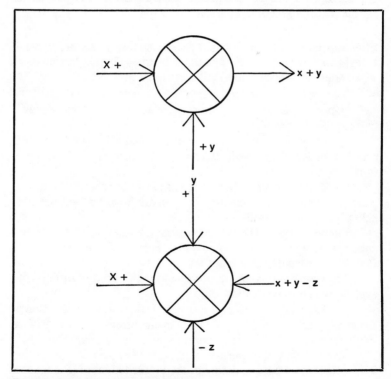

Fig. 2-3. Summary point representation.

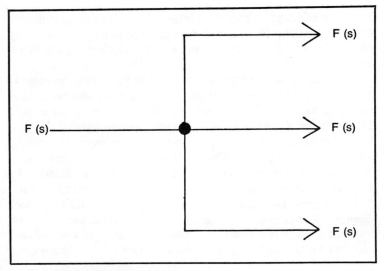

Fig. 2-4. Take-off point representation.

tities represent time functions. Upper case letters denote Laplace transformed quantities as functions of the complex variable, s, when these quanities represent frequency functions.

A *controller* is also called the *control elements* (G_1). These are the components required to generate the appropriate control signal (M) applied to the plant (G_2).

A *plant* is also called the *controlled system* (G_2). It is the body, process, or machine, of which a particular quantity or condition is to be controlled.

Feedback elements (H) are the components required to establish the functional relationship between the primary feedback signal (B) and the controlled output (C).

Reference input is the external signal (R) applied to a feedback control system in order to command a specified action of the plant. It often represents ideal plant output behavior.

The controlled output is the quantity (C) or condition of the system which is controlled.

A *primary feedback signal* is a signal (B) which is a function of the controlled output (C), and which is algebraically summed with the reference input (R) to obtain the actuating signal (E).

An *error* also called an *actuating signal* (E) or *control action*. It is the algebraic sum consisting of the reference input (R), plus or minus the primary feedback (B).

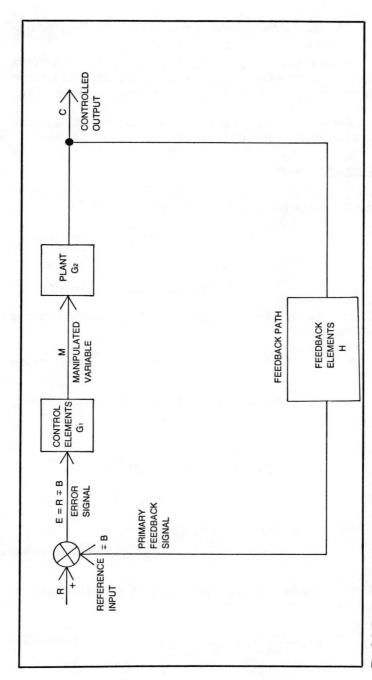

Fig. 2-5. Block diagram of a feedback control system.

The *manipulated variable* is that quantity (M) which the control elements (G_1) apply to the plant (G_2).

A *disturbance* is an undesired input signal (U) which affects the value of the controlled output (C). It may enter the plant by summation with M, or via an intermediate point, as shown in Fig. 2-5.

Forward path is the transmission path from the actuating signal (E) to the controlled output (C).

FEEDBACK CONTROL SYSTEM

The closed loop transfer function, $T(s) = C(s)/R(s)$, can be expressed as a function of G(s) and H(s) from Fig. 2-6.

$$C(s) = G(s)E(s) \qquad \textbf{Equation 2-1}$$

$$B(s) = H(s)C(s) \qquad \textbf{Equation 2-2}$$

$$E(s) = R(s) - B(s) \qquad \textbf{Equation 2-3}$$

Substituting Equation 2-3 into Equation 2-1 yields the following results.

$$C(s) = G(s) R(s) - G(s)B(s) \qquad \textbf{Equation 2-4}$$

Substituting equation 2-2 into equation 2-4, we have the following result.

$$C(s) = G(s) R(s) - G(s)H(s)C(s)$$

$$G(s)R(s) = C(s) + C(s)G(s)H(s)$$

$$\frac{G(s)R(s)}{C(s)} = 1 + G(s)H(s) \qquad \textbf{Equation 2-5}$$

Then solving the above equation for $T(s) = C(s)/R(s)$, we have the following result.

$$T(s) = \frac{C(s)}{R(s)} = \frac{G(s)}{1 + G(s)H(s)} \qquad \textbf{Equation 2-6}$$

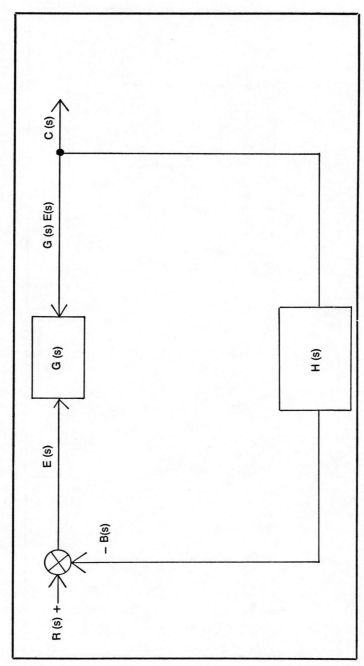

Fig. 2-6. Basic block diagram of a feedback control system.

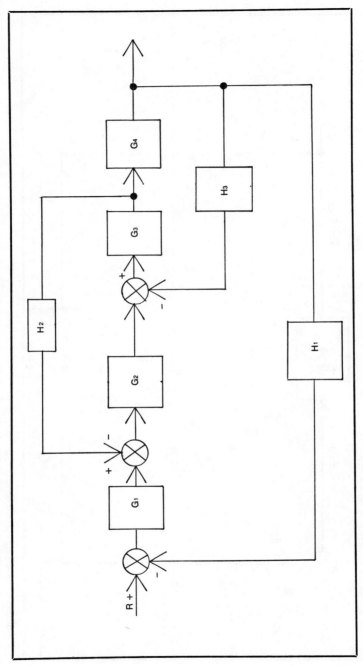

Fig. 2-7. Multiple feedback loop control system.

32

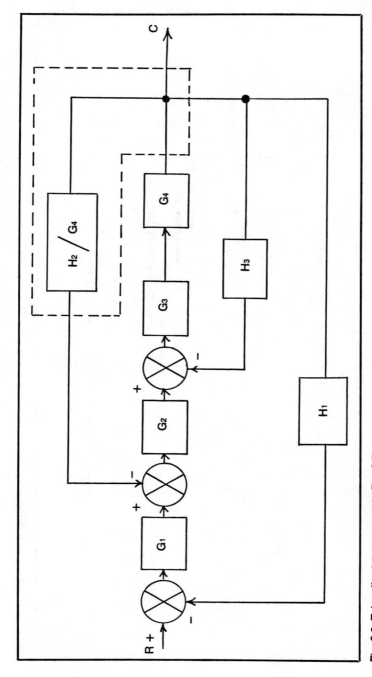

Fig. 2-8. Take-off point movement in Fig. 2-7.

33

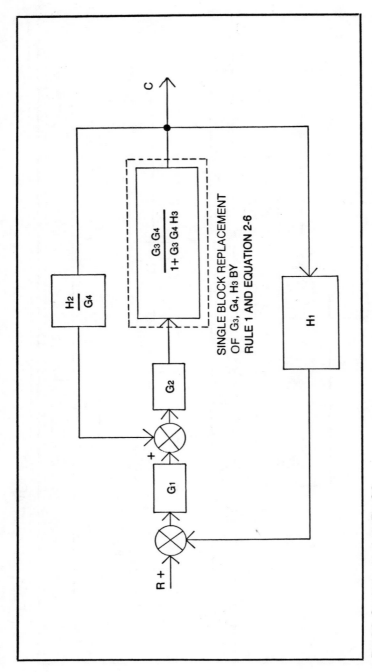

C

$$\frac{G_3 \, G_4}{1+ G_3 \, G_4 \, H_3}$$

SINGLE BLOCK REPLACEMENT
OF G_3, G_4, H_3 BY
RULE 1 AND EQUATION 2-6

$\dfrac{H_2}{G_4}$

G_2

G_1

H_1

R +

+

Fig. 2-9. Combining blocks in Fig. 2-8.

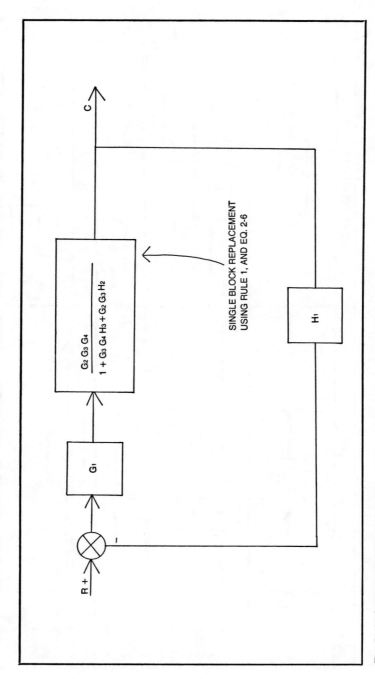

$$\frac{G_2 G_3 G_4}{1 + G_3 G_4 H_3 + G_2 G_3 H_2}$$

SINGLE BLOCK REPLACEMENT
USING RULE 1, AND EQ. 2-6

G_1

H_1

R +

—

C

Fig. 2-10. Combining blocks in Fig. 2-9.

35

Table 2-1. Block Diagram Manipulations.

Rules	ORIGINAL NETWORK	EQUIPMENT NETWORK
1. Cascaded Elements	R → [G₁] → [G₂] → C	R → [G₁ G₂] → C
2. Addition and Subtraction	R → [G₁] → ⊗(+, ±) → C, [G₂]	R → [G₁ ± G₂] → C
3. Moving a starting point in Front of an element	R → [G] → • → C	R → • → [G] → C, • → [G]
4. Moving a starting point behind an element	R → [G] → C (with branch)	R → [G] → • → C, [1/G]

Table 2-1. Block Diagram Manipulations.

Rules	ORIGINAL NETWORK	EQUIVALENT NETWORK
5. Moving a Summing point ahead of an element		
6. Moving a Summing point behind an element		

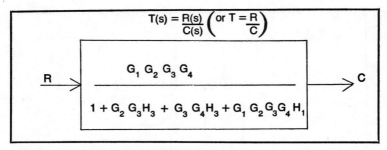

Fig. 2-11. Final block reduction of Fig. 2-7.

The block diagram of complicated feedback control systems usually contains several feedback loops, and the evaluation of the system transfer functions directly from the block diagrams is usually tedious. The transfer function of a complex block diagram configuration is obtained by using the block diagram reduction technique. The block diagram of the system is first reduced to the basic form in Fig. 2-6 and then the transfer function is written from Equation 2-6. Some of the important block diagram reduction manipulations are given in the Table 2-1.

The block diagram of a multiple feedback loop system is shown in Fig. 2-7. The closed loop transfer function of the system, C(s)/R(s), is developed through the block diagram reduction technique with the aid of the rules in Table 2-1. Figures 2-9 through 2-12 illustrate the reduction technique via block diagrams.

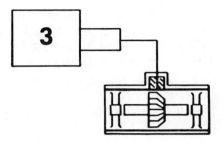

The Mathematics of Control Systems

Control systems can be defined mathematically by ordinary differential equations with constant coefficients. However, the differential equations can be converted to the frequency domain through a branch of mathematics called *Laplace transforms*. Under the conditions that the necessary transforms exist, the Laplace transform method consists of three basic steps:

☐ Take the Laplace transform of each side of a differential equation and equate to obtain an algebraic equation in the transform of the unknown function.

☐ Solve the algebraic equation obtained in the previous step for the transform of the unknown function.

☐ Find the inverse transform which will be the time solution to the original differential equation.

The method just described is analogous to the method of logarithms in multiplication. For example, finding the product of two numbers, say 3.1416 times 7.2314, by employing logarithms can be diagramed as in Fig. 3-1. The Laplace transform method of solving a differential equation with initial conditions included is shown in Fig. 3-2.

We will not derive the Laplace transforms of any time functions, but we will employ the Laplace transform table in Appendix B much as a common logarithm table in order to solve the differential equations presented throughout this chapter and later in this book.

□ **Example 3-1:** Find the solution for v(t) of the differential equation,

$$\frac{d^2v(t)}{dt} - 4\frac{dv(t)}{dt} + 4\,v(t) = 0,$$

with the initial conditions, $v(0) = 0$ and $\dfrac{dv(0)}{dt} = 1.$

After finding the Laplace transform of the first and second derivatives from the transform table in Appendix A, the differential time equation given in this problem is transformed into the *s* domain or frequency domain by the following equation.

$$[s^2\,V(s) - s\,v(0) - v'(0)] - 4s\,V(s) + 4\,v(0) + [4\,V(s)] = 0$$

Next, the initial conditions are substituted into the previous equation.

$$s^2\,V(s) - s(0) - (1) - 4s\,V(s) + 4(0) + 4\,V(s) = 0$$

Collecting terms that contain V(s) yields the following equation.

$$(s^2 - 4s + 4)V(s) = 1$$

This equation is then solved for V(s).

$$V(s) = \frac{1}{s^2 - 4s + 4}$$

Next, the roots and the factors of the denominator of the above equation are found, which are $s_1 = -2$, and $s_2 = -2$. Then the above equation can be written in terms of the factors as follows.

$$V(s) = \frac{1}{(s - 2)^2}$$

The preceding step matches the V(s) equation with a transform pair in the table of Appendix A. If you look in Appendix A, you will find the desired time solution to be given by the following equation.

$$v(t) = L^{-1}\,V(s) = t\,e^{2t}$$

□ **Example 3-2:** Find the solution for v(t) of the differential equation $x''(t) + 4\,x(t) = 0$, with the following initial conditions: $x(0) = 1$, and $x(\pi/4) = -1$. Note, that $x''(t)$ means the second

40

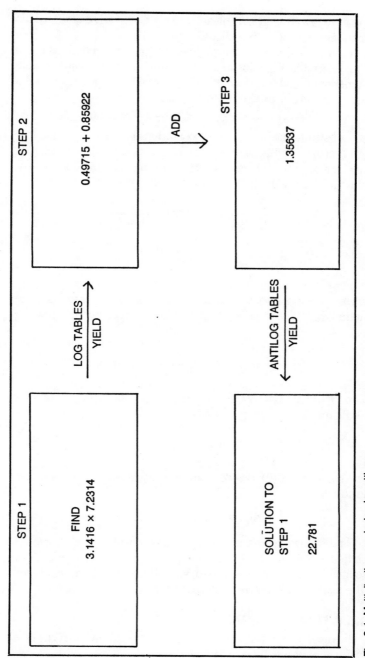

Fig. 3-1. Multiplication employing logarithms.

derivative of x(t), and x'(t) means the first derivative of x(t). This symbolism is used in the chapters that follow.

Find the Laplace transform pairs for the time functions given in the differential equation of this example from the transform table in Appendix A and transform the differential time equation into the frequency domain by the following equation.

$$s^2 X(s) - s\, x(0) - x'(0) + 4\, X(s) = 0$$

Next substitute the known initial conditions in the above equation; that is, $x(0) = 1$. Because the value of $x'(0)$ is unknown, carry it along as an arbitrary constant which will be determined later. At this point, solve for X(s) as shown.

$$X(s) = \frac{s}{s^2 + 4} + x'(0) \left(\frac{1}{s^2 + 4} \right)$$

$$= \frac{s}{s^2 + 4} + \left(\frac{x'(0)}{2} \right) \left(\frac{2}{s^2 + 4} \right)$$

Look in Appendix A to find the desired time solution for the Laplace equation X(s). The resulting time solution is given by the following equation.

$$x(t) = \cos 2t + \frac{x'(0)}{2} \sin 2t$$

To find $x'(0)$, let $t = \pi/4$ and use the condition $x(\pi/4) = -1$. You will obtain the following equation.

$$-1 = 0 + \frac{x'(0)}{2} \cdot 1$$

Then solve for $x'(0) = -2$ from this equation. Therefore, the solution is:

$$x(t) = \cos 2t - \sin 2t$$

The following sections of this chapter will deal with special techniques to find time solutions from differential equations through Laplace transforms. Also, special time functions employed in the analysis of control systems will also be covered.

PARTIAL FRACTION EXPANSION

In most control system problems, the following form of a transfer function will be obtained.

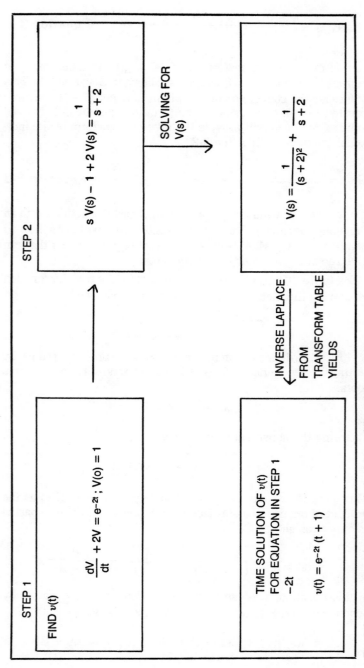

STEP 1

FIND $v(t)$

$$\frac{dV}{dt} + 2V = e^{-2t} \; ; V(o) = 1$$

STEP 2

$$s V(s) - 1 + 2 V(s) = \frac{1}{s + 2}$$

SOLVING FOR V(s)

$$V(s) = \frac{1}{(s + 2)^2} + \frac{1}{s + 2}$$

INVERSE LAPLACE

FROM TRANSFORM TABLE YIELDS

TIME SOLUTION OF $v(t)$ FOR EQUATION IN STEP 1

$$v(t) = e^{-2t} (t + 1)$$

Fig. 3-2. Laplace transform method of solving a differential equation.

$$G(s) = \frac{1}{s^n + a_1 s^{n-1} + a_2 s^{n-2} + \ldots + a_n} \quad \text{Equation 3-1}$$

To find the inverse transform, $L^{-1}G(s) = g(t)$, it is necessary to express G(s) as a sum of fractions, as shown in Equation 3-2. This is because the conversion tables will usually not give the inverse transform of an expression of the form in Equation 3-1, but the conversion tables will provide the inverse transform of the individual fractions in Equation 3-2.

$$G(s) = \frac{K_1}{s - s_1} + \frac{K_2}{s - s_2} + \ldots + \frac{K_n}{s - s_n} \quad \text{Equation 3-2}$$

We will now illustrate the method of partial fractions with an elementary example. Later in this section, we will show how to evaluate $L^{-1}G(s)$ when G(s) is a quotient of polynomials that can have repeating roots or imaginary roots.

□**Example 3-3:** Find the solution for g(t) for the transfer function employing the method of partial fraction expansion.

$$G(s) = \frac{1}{s^2 - s - 2,}$$

First, write the denominator of G(s) as the product of sums, by setting the denominator of G(s) equal to zero and solving for its two roots.

$$s^2 - s - 2 = 0$$

Factoring the above equation yields:

$$(s + 1)(s - 2) = 0$$

This means the roots are $s_1 = -1$ and $s_2 = 2$. Write G(s) as the product of the denominator factors and then split G(s) into partial fractions, as shown.

$$G(s) = \frac{1}{(s + 1)(s - 2)} = \frac{A}{s + 1} + \frac{B}{s - 2}$$

In this equation, you can multiple both side of the equation by denominator $(s + 1)(s - 2)$ to obtain the following result.

$$1 = A(s - 2) + B(s + 1) = s(A + B) + (-2A + B)$$

To determine the constants A and B, equate coefficients of like powers of s on both sides of the previous equation. Hence, you can write the following equations.

$$0s = (A + B)s$$
or
$$0 = A + B$$
$$1 = -2A + B$$

In the first equation, $A = -B$, which is substituted into the second equation to obtain:

$$1 = -2(-B) + B = 3B$$

Hence, solving for B yields $B = \frac{1}{3}$, which means that $A = -\frac{1}{3}$.

Another method of solving for the constants A and B makes use of roots $s_1 = -1$ and $s_2 = 2$. Start with the equation:

$$1 = A(s - 2) + B(s + 1) \qquad \textbf{Equation 3-3}$$

Substitute $s_1 = -1$ into this equation and solve for A.

$$1 = A(-1 - 2) + B(-1 + 1) = A(-3) + B(0)$$
$$A = -\frac{1}{3}$$

Substitute $s_2 = 2$ into Equation 3-3 and solve for B.

$$1 = A(2 - 2) + B(2 + 1) = A(0) + B(3)$$
$$B = \frac{1}{3}$$

The same results are obtained by either method. Although the second method is easier at this time, the first method will be employed for the most part as the transfer function G(s) becomes more complex. You should know both methods.

G(s) can be written in the partial fraction form that follows:

$$G(s) = \frac{-\frac{1}{3}}{s + 1} + \frac{\frac{1}{3}}{s - 2}$$

Look in Appendix A to find the desired time solution for this Laplace equation, G(s). The transform number (No.) in Appendix A is No. 3, and the time solution is written as follows:

$$g(t) = \frac{-1}{3}\, e^{-t} + \frac{1}{3}\, e^{2t}$$

To summarize this example, a partial fraction expansion of $G(s)$ is obtainable when the numerator of $G(s)$ is a real number, and no double or repeating roots exist in the denominator of $G(s)$ (i.e., when $s_1 = s_2 = 4$) as follows:

☐ Find the roots s_1, s_2, . . ., etc. of the denominator of $G(s)$.
☐ Write $G(s)$ in the form:

$$G(s) = \frac{K}{(s^n + a_1 s^{n-1} + \ldots + a_n s^0)} = \frac{A_1}{(s - s_1)} + \cdots + \frac{A_n}{(s - s_n)}$$

☐ Sum this equation over a common denominator and then find the constants A_1, A_2, . . ., A_n by setting $s = s_1$ and obtaining an equation, $s = s_2$, and an equation all the way through $s = s_n$. Enough equations are obtained to find all of the constants in the partial fraction expansion. Remember that this procedure is useful only when the numerator is a number and no repeating roots exist in the denominator of $G(s)$.

If $N(s)$ and $P(s)$ are polynomials in s, with real coefficients and if the degree of $N(s)$ is less than that of $P(s)$, the inverse Laplace transform $L^{-1} N(s)/P(s)$ exists and is often evaluated by use of a partial fraction decomposition of the fraction $N(s)/P(s)$. We assume the possibility of this partial fraction decomposition in the domain of real numbers and suggest various techniques for obtaining it. In the cases usually encountered, the existence of $L^{-1} N(s)/P(s)$ will be apparent from the partial fraction decomposition of the fraction, $N(s)/P(s)$. We shall also consider briefly the partial fraction decomposition in the domain of complex numbers, in which $P(s)$ is factored completely into linear factors and from which the existence of $L^{-1} N(s)/P(s)$ follows in general.

Recall from elementary algebra that apart from a constant factor, $P(s)$ can be factored into a product of distinct real factors of two types: $(s - r)^p$ and $(s^2 + cs + d)^q$, where p and q are positive integers, and r is a real root of $P(s) = 0$. The theory of partial fraction decomposition states that the fraction, $N(s)/P(s)$, is the sum of two types of sums. The first type corresponds to a factor, $(s - r)^p$, and has the form of Equation 3-4 shown below:

$$\frac{A_1}{(s - r)} + \frac{A_2}{(s - r)^2} + \cdots + \frac{A_p}{(s - r)^p} \qquad \textbf{Equation 3-4}$$

where A_1, A_2, . . ., A_p are constants.

The second type corresponds to a factor, $(s^2 + cs + d)$, and has the form:

Equation 3-5

$$\frac{C_1s + D_1}{s^2 + cs + d} + \frac{C_2s + D_2}{(s^2 + cs + d)^2} + \cdots + \frac{C_qs + D_q}{(s^2 + cs + d)^q}$$

where C_1, C_2, . . ., C_q and D_1, D_2, . . ., D_q are constants. It is sometimes convenient to express $s^2 + cs + d$ as $(s - a)^2 + b^2$, by employing the method of completing the square in algebra.

The simplest case of partial fraction decomposition occurs for the fraction $N(s)/P(s)$, when all the roots of $P(s) = 0$ are real and distinct. If these roots are called s_1, s_2, . . ., s_n, constants A_1, A_2, . . ., A_n must be found such that the following equation can be written.

$$\frac{N(s)}{P(s)} = \frac{A_1}{s - s_1} + \frac{A_2}{s - s_2} + \cdots + \frac{A_n}{s - s_r} \quad \textbf{Equation 3-6}$$

The following is a quick method for the evaluation of A_1, A_2, . . ., A_n. Define the quantity $B_i(s)$:

$$B_i(s) = (s - s_i)\ \frac{N(s)}{P(s)}\ , (i = 1, 2, \ldots, n) \quad \textbf{Equation 3-7}$$

where $(s - s_i)$ cancels into the same factor in the denominator term $P(s)$.

If $s = s_i$ is substituted into Equation 3-7 after the $(s - s_i)$ terms are canceled, the following is obtained:

$$B_i(s_i) = A_i \quad \textbf{Equation 3-8}$$

☐ **Example 3-4:** Find the partial fraction expansion of the function $\dfrac{s^3 + 1}{3(s - 1)\,(s + 2)\,(s + 3)\,(s - 5)}$. Then use the expansion to find the inverse transform of the given function. First set up the given fraction $N(s)/P(s)$ as follows;

$$\frac{N(s)}{P(s)} = \frac{A_1}{s - 1} + \frac{A_2}{s + 2} + \frac{A_3}{s + 3} + \frac{A_4}{s - 5}$$

Next, find each of the B_i (s) terms by employing Equation 3-7.

$$B_1(s) = \frac{(s - 1)(s^3 + 1)}{3(s - 1)(s + 2)(s + 3)(s - 5)}$$

$$B_1(s) = \frac{s^3 + 1}{3(s + 2)(s + 3)(s - 5)}$$

$$B_2(s) = \frac{s^3 + 1}{3(s - 1)(s + 3)(s - 5)}$$

$$B_3(s) = \frac{s^3 + 1}{3(s - 1)(s + 2)(s - 5)}$$

$$B_4(s) = \frac{s^3 + 1}{3(s - 1)(s + 2)(s + 3)}$$

Now find the constants by employing Equation 3-8.

$$A_1 = B_1(1) = \frac{1^3 + 1}{3(1 + 2)(1 + 3)(1 - 5)} = \frac{-1}{72}$$

$$A_2 = B_2(-2) = \frac{(-2)^3 + 1}{3(-2 - 1)(-2 + 3)(-2 - 5)} = \frac{-1}{9}$$

$$A_3 = B_3(-3) = \frac{(-3)^3 + 1}{3(-3 - 1)(-3 + 2)(-3 - 5)} = \frac{13}{48}$$

$$A_4 = B_4(5) = \frac{(5)^3 + 1}{3(5 - 1)(5 + 2)(5 + 3)} = \frac{3}{16}$$

Therefore, the given function can be expressed by the following partial fraction expansion:

$$\frac{N(s)}{P(s)} = \frac{A_1}{s - 1} + \frac{A_2}{s + 2} + \frac{A_3}{s + 3} + \frac{A_4}{s - 5}$$

To find the inverse Laplace transform of the previous equation, refer to Appendix A, transform No. 3, and the time solution is written as follows.

$$L^{-1}\frac{N(s)}{P(s)} = \frac{-1}{72}e^t + \frac{-1}{9}e^{-2t} + \frac{13}{48}e^{-3t} + \frac{3}{16}e^{5t}$$

The next example of partial fraction expansion is when we encounter repeating roots of s in the denominator term, $P(s)$.

□ **Example 3-5:** Find the partial fraction expansion and the inverse Laplace transform of the equation, $\dfrac{s^2}{(s + 1)^3}$

First set up the given fraction, $N(s)/P(s)$, as follows:

$$\frac{N(s)}{P(s)} = \frac{A_1}{(s + 1)} + \frac{A_2}{(s + 1)^2} + \frac{A_3}{(s + 1)^3}$$

In this equation, you can multiply both sides of the equation by the denominator $(s + 1)^3$ to obtain the following results.

$$s^2 = A_1(s + 1)^2 + A_2(s + 1) + A_3$$

To determine the above constants, equate coefficients of like powers of s on both sides of the equation. Hence, the following equations can be written.

$$s^2 = A_1(s^2 + 2s + 1) + A_2(s + 1) + A_3$$

$$s^2 = (A_1)s^2 + (2A_1 + A_2)s + (A_1 + A_2 + A_3)$$

$$1 = A_1$$

$$0 = 2A_1 + A_2$$

$$0 = A_1 + A_2 + A_3$$

Substituting $A_1 = 1$ into the second of these equations yields:

$$0 = 2(1) + A_2$$

$$-2 = A_2$$

Substituting the values of A_1 and A_2 into the third equation yields:

$$0 = 1 + (-2) + A_3$$

$$1 = A_3$$

Therefore, the given function can be expressed by the following partial fraction expansion:

$$\frac{N(s)}{P(s)} = \frac{1}{s+1} + \frac{-2}{(s+1)^2} + \frac{1}{(s+1)^3}$$

To find the inverse Laplace transform of the above equation, refer to Appendix A and apply transform No. 3 to the first fraction, and apply transform No. 4 to the second and third fractions, which results in the following time solution:

$$L^{-1}\frac{N(s)}{P(s)} = e^{-t} + \frac{(-2)}{(2-1)!}t^{(2-1)}e^{-2t} + \frac{1}{(3-1)!}t^{(3-1)}e^{-3t}$$

$$= e^{-t} + \frac{(-2)}{1!}t\,e^{-2t} + \frac{1}{2!}t^2\,e^{-3t}$$

$$= e^{-t} - 2t\,e^{-2t} + \frac{1}{(2)\,(1)}t^2\,e^{-3t}$$

$$= e^{-t} - 2t\,e^{-2t} + 0.5\,t^2\,e^{-3t}$$

Next, consider the problem of finding the inverse transform of a fraction, $N(s)/P(s)$, in which denominator $P(s)$ is a simple mixture of an *unrepeated linear factor* and an *unrepeated quadratic factor*.

□**Example 3-6:** Find the partial fraction expansion and the inverse Laplace transform of the equation $\dfrac{s+1}{s(s^2+1)}$.

First set up the given fraction, $N(s)/P(s)$, as follows:

$$\frac{N(s)}{P(s)} = \frac{A}{s} + \frac{Bs+C}{s^2+1} \qquad \textbf{Equation 3-9}$$

Solve for the roots of the denominator by letting $P(s) = 0$.

$$0 = s(s^2+1)$$

$$0 = s_1$$

$$-1 = s^2$$

$$\left.\begin{array}{l} +i = s_2 \\[1.5em] -i = s_3 \end{array}\right\} \text{ imaginary roots, that are complex conjugate}$$

Next, multiply Equation 3-9 by s and substitute $s_1 = 0$ into the resulting equation.

$$\frac{N(s)}{P(s)}\, s = \frac{A}{s}\, s + \frac{Bs^2 + Cs}{s^2 + 1} = \frac{(s+1)\, s}{s(s^2 + 1)}$$

Substituting $s_1 = 0$ into this equation yields:

$$A + \frac{B(0)^2 + C(0)}{(0)^2 + 1} = \frac{(0+1)}{(0^2 + 1)}$$

$$A = 1$$

Multiply equation 3-9 by $(s^2 + 1)$ to obtain the following result:

$$\frac{A(s^2 + 1)}{s} + \frac{(Bs + C)(s^2 + 1)}{s^2 + 1} = \frac{(s+1)(s^2 + 1)}{s(s^2 + 1)}$$

$$\frac{A(s^2 + 1)}{s} + Bs + C = \frac{s+1}{s}$$

Substitute $s_2 = i$ in the previous equation to obtain the following result:

$$\frac{A(i^2 + 1)}{i} + Bi + C = \frac{i+1}{i}$$

$$\frac{A(-1 + 1)}{i} + Bi + C = \frac{i+1}{i}$$

$$0 + (Bi + C)i = i + 1$$

$$Bi^2 + Ci = i + 1$$

$$-B + Ci = i + 1$$

Equating like coefficients from both sides of the above equation yields the following results.

$$-B = 1$$

$$B = -1$$

$$Ci = i$$

$$C = 1$$

Therefore, the given function can be expressed by the following partial fraction expansion:

$$\frac{N(s)}{P(s)} = \frac{1}{s} + \frac{-s + 1}{s^2 + 1}$$

The second fraction of this equation can be separated into two fractions, and the equation can be written as follows.

$$\frac{N(s)}{P(s)} = \frac{1}{s} - \frac{s}{s^2 + 1} + \frac{1}{s^2 + 1}$$

To find the inverse Laplace transform of the above equation, refer to Appendix A and apply transform No. 18 to the first fraction, transform No. 11 to the second fraction, and transform No. 10 to the third fraction. This results in the following time solution:

$$f(t) = 1 - \cos t + \sin t$$

For our next example, consider a fraction whose denominator has a *single quadratic factor* and a *repeated linear factor*.

☐ **Example 3-7:** Find the partial fraction expansion and the inverse Laplace transform of the equation, $\dfrac{s + 1}{s^2(s^2 + s + 1)}$.

First we set up the given equation as follows:

$$\frac{N(s)}{P(s)} = \frac{A}{s} + \frac{B}{s^2} + \frac{Cs + D}{s^2 + s + 1} \qquad \textbf{Equation 3-10}$$

Solve for the roots of the denominator by letting $P(s) = 0$.

$0 = s^2$

$0 = s_1$ a repeating root since the power of s is 2

$0 = s^2 + s + 1$

Applying the quadratic equation to the above equation clearly defines two roots which are the complex conjugate of each other.

These roots are found to be $s_2, s_3 = -0.5 \pm \iota 0.5\sqrt{3}$. You can obtain A and independently of C and D by first multiplying Equation 3-10 by s^2 and obtaining the following result:

$$\frac{(s + 1)s^2}{s^2(s^2 + s + 1)} = \frac{As^2}{s} + \frac{Bs^2}{s^2} + \frac{(Cs + D)s^2}{s^2 + s + 1}$$

Equation 3-11

Letting $s = 0$ obtains the following result. Remember to cancel the s^2 terms before letting $s = 0$.

$$\frac{0 + 1}{0^2 + 0 + 1} = A(0) + B0 + \frac{(C(0) + D^2}{0^2 + 0 + 1}$$

$$1 = B$$

If you attempt to obtain A by multiplying Equation 3-10 by s and then set $s = 0$, two fractions of the form $1/0$, which are undefined, will be encountered. Hence, L'Hospital's rule must be applied. Take the derivative of Equation 3-11 and set $s = 0$ to evaluate A. When you differentiate Equation 3-11 with respect to s, you obtain the following result.

$$\frac{s^2 + s + 1 - (s + 1)(2s + 1)}{(s^2 + s + 1)^2} = A + \frac{(s^2 + s + 1)ds^2)(2s + 1}{(s^2 + s + 1)^2}$$

Setting $s = 0$ results in the following equation:

$$0 = A + 0$$

Clearly, for $s = 0$, $A = 0$ and $B = 1$.

To compute C and D, multiply Equation 3-10 by $(s^2 + s + 1)$ and obtain the following result.

$$\frac{s + 1}{s^2} = \frac{A(s^2 + s + 1)}{s} + \frac{B s^2 + s + 1)}{s^2} + Cs + D$$

Setting s equal to a root of $s^2 + s + 1 = 0$ amounts to using $s^2 = -s - 1$ in the last equation. When this is substituted into the equation, the following results:

$$-1 = Cs + D$$

Substituting one root of $s^2 + s + 1 = 0$ into the equation yield the following result:

$$-1 = C(-0.5 + i0.5\ 3) + D$$

Equating like coefficients from both sides of the equation yields the following results:

$$D = -1$$

$$C = 0$$

You can express the given function by the following partial fraction expansion:

$$\frac{N(s)}{P(s)} = \frac{1}{s^2} - \frac{1}{s^2 + s + 1}$$

To find the inverse Laplace transform of this equation, refer to Appendix A and apply transform No. 29 to the first fraction and transform No. 15 to the second fraction. The inverse transform of the first fraction is very simple; that is, the answer is t. However, it is necessary to perform the following calculations to find the inverse transform of the second fraction.

$$\frac{1}{s^2 + s + 1} = \frac{1}{s^2 + 2\zeta\omega_n s + \omega_n^2}$$

Comparing denominator terms of like powers of s, the following calculations can be made.

$$\omega_n^2 = 1; \text{ therefore } \omega_n = 1$$

$$2\zeta\omega_n = 1$$

$$\zeta = \frac{1}{2\omega_n} = \frac{1}{2(1)} = 0.5$$

$$\omega_d = \omega_n \sqrt{1 - \zeta^2} = (1)\sqrt{1 - (0.5)^2}$$

$$\omega_d = 0.866$$

At this time, the time solution can be written using the the results just tabulated.

$$f(t) = t - \frac{1}{0.866} \, e^{-0.5t} \sin 0.866t$$

$$f(t) = t - 1.1547 \, e^{-0.5t} \sin 0.866t$$

The next two examples will illustrate special cases of the previous seven examples, but they will serve to guide the reader in the quest for partial fraction expansion.

□ **Example 3-8:** Find the partial fraction expansion and the inverse Laplace transform of the equation, $\dfrac{2s}{(s^2 + 1)(s^2 + 2)}$

Set up the given equation as follows:

$$\frac{2s}{(s^2 + 1)(s^2 + 2)} = \frac{As + B}{s^2 + 1} + \frac{Cs + D}{s^2 + 2} \quad \textbf{Equation 3-12}$$

If we multiply equation 3-12 by $s^2 + 1$, the following result is obtained:

$$\frac{2s}{s^2 + 2} = As + B + (s^2 + 1) \frac{(Cs + D)}{s^2 + 2} \quad \textbf{Equation 3-13}$$

If Equation 3-12 is multiplied by $s^2 + 2$, the following result is obtained:

$$\frac{2s}{s^2 + 1} = (s^2 + 2) \frac{(As + B)}{s^2 + 1} + Cs + D \quad \textbf{Equation 3-14}$$

Next find the roots of $(s^2 + 1)$ and $(s^2 + 2)$.

$$s^2 + 1 = 0$$
$$s_1, s_2 = \pm i$$

$$s^2 + 2 = 0$$

$$s_3, s_4 = \pm i(1.414)$$

Denote the left sides of Equation 3-13 and Equation 3-14 by the following equations:

$$\phi_1(s) = \frac{2s}{s^2 + 2}$$

$$\phi_2(s) = \frac{2s}{s^2 + 1}$$

These yield the following:

$$\phi_1(\iota) \qquad = 2i = Ai + B$$

$$\phi_1(\iota\, 1.414) = -2i(1.413) = C(i1.414) + D$$

These two equations may be solved for A, B, C, and D by equating real and imaginary parts. Thus, you will find the following: A = 1, B = 0, C = −2, and D = 0.

Now, you can express the given function by the following partial fraction expansion:

$$\frac{N(s)}{P(s)} = \frac{2s}{s^2 + 1} - \frac{2s}{s^2 + 2}$$

To find the inverse Laplace transform of this equation, refer to Appendix A and apply transform No. 11 to the first fraction and the second fraction. The resulting time solution is given as follows.

$$f(t) = 2 \cos t - 2 \cos 1.414\, t$$

□**Example 3-9:** Find the partial fraction expansion and the inverse Laplace transform of the equation, $\dfrac{2s^3 + 1}{(s^2 + 4)^2}$

Set up the equation in the following form:

$$\frac{2s^3 + 1}{(s^2 + 4)^2} = \frac{As + B}{s^2 + 4} + \frac{Cs + D}{(s^2 + 4)^2} \qquad \textbf{Equation 3-15}$$

Next, multiply both sides of Equation 3-15 by the highest power of the repeated factor to obtain the following equation:

$$2s^2 + 1 = (As + B)(s^2 + 4) + Cs + D$$

The roots of the denominator are +2i and −2i. Substitute s = 2i into the previous equation to obtain the following result:

$$2(2i)^3 + 1 \;=\; \left[A(2i) + B \right] \left[(2i)^2 + 4 \right] + Cs + D$$

$$-16i + 1 \;=\; (2Ai + B)(-4 + 4) + c(2i) + D$$

$$-16i + 1 \;=\; 2Ci + D$$

This equation can be solved for C and D by equating coefficients of real and imaginary parts. Hence, $D = 1$ and $C = -8$. Next, substitute the values of C and D into Equation 3-15 and transpose the last fraction in Equation 3-15 from the right side of the equation to the left side of the equation to obtain the following result:

$$\frac{2s^3 + 1}{(s^2 + 4)^2} - \frac{-8s + 1}{(s^2 + 4)^2} = \frac{As + B}{s^2 + 4}$$

Solving this equation results in the following:

$$2s = As + B$$

It is now obvious that $B = 0$ and $A = 2$ if the previous equation is solved.

You can express the given function by the following partial fraction expansion:

$$\frac{N(s)}{P(s)} = \frac{2s}{s^2 + 4} + \frac{-8s + 1}{(s^2 + 4)^2}$$

The second fraction of this equation can be separated into two fractions, and the equation can be written as follows.

$$\frac{N(s)}{P(s)} = \frac{2s}{s^2 + 4} + \frac{-8s}{(s^2 + 4)^2} + \frac{1}{(s^2 + 4)^2}$$

To find the inverse Laplace transform of this equation, refer co Appendix A and apply transform No. 11 to the first fraction, transform No. 37 to the second fraction, and transofrm No. 10 to the third fraction. The second and third fraction need a slight adjustment to match up with the transforms numbered 37 and 10. The adjustments follow.

Second Fraction:

$$\frac{-8s}{(s^2 + 4)^2} = (-2)\frac{(2s)(2)}{(s^2 + 4)^2} = (-2)\frac{2s\omega}{(s^2 + \omega^2)^2}$$

Third Fraction:

$$\frac{1}{(s^2 + 4)^2} = \left(\frac{1}{2}\right)\frac{2}{(s^2 + 4)^2} = (0.5)\frac{\omega}{(s^2 + \omega^2)^2}$$

The time function that results from both adjustments is the following:

$$f(t) = 2\cos 2t - 2t\sin 2t + 0.5\sin 2t$$

THE UNIT STEP FUNCTION AND PERIODIC FUNCTIONS

One of the simplest discontinuous (abrupt change) functions is the *unit step* function, which is defined mathematically as follows:

$$\left. \begin{array}{l} u(t - t_o) = 0, \text{ for } t < t_o \\ u(t - t_o) = 1, \text{ for } t \geq t_o \end{array} \right\} \quad \textbf{Equation 3-16}$$

Note that u is a function of t and that t_o plays the role of a parameter which indicates the point at which a unit step occurs. The graph of the unit step defined by Equation 3-16 is shown in Fig. 3-3.

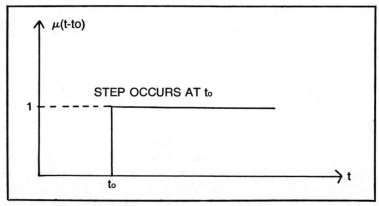

Fig. 3-3. Unit step functions.

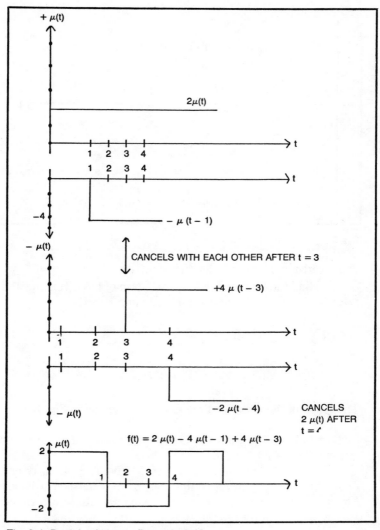

Fig. 3-4. Partial solution to Example 3-10.

The Laplace transform of the unit step function is given in Appendix A transform No. 18, which is the step function occurring at $t_o = 0$. The transform of the delayed unit step function, as shown in Fig. 3-3, has as its transform No. 19, where $t_o = T$.

☐**Example 3-10:** Find the inverse Laplace transform of the function, $G(s) = \dfrac{2}{s}$ $(1 - 2e^{-s} + 2e^{-3s} - e^{-4s})$,

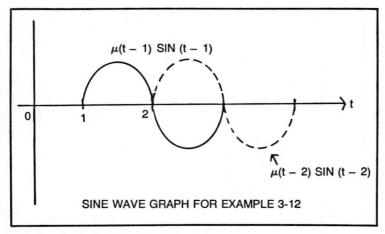

$\mu(t - 1)$ SIN $(t - 1)$

$\mu(t - 2)$ SIN $(t - 2)$

SINE WAVE GRAPH FOR EXAMPLE 3-12

Fig. 3-5. Sine wave for Example 3-12.

and sketch a graph of the time function.

To find the inverse Laplace transform of the given equation, refer to Appendix A and apply No. 18 and No. 19.

$$L^{-1}G(s) \;=\; \frac{L^{-1}\,2}{s} \;-\; \frac{L^{-1}\,4e^{-s}}{s} \;+\; \frac{L^{-1}\,4e^{-3s}}{s} \;-\; \frac{L^{-1}\,2e^{-4s}}{s}$$

$$f(t) = 2u(t) - 4u(t - 1) + 4u(t - 3) - 2u(t - 4)$$

The graph of the previous time function is shown in Fig. 3-4.

☐ **Example 3-11:** Find the Laplace transform of the function, $g(t) = u(t - 2) \sin (t - 2)$.

To find the Laplace transform of the given equation, refer to Appendix B and apply operation No. 8. You must also look in Appendix A and use transform No. 19, along with transform No. 10.

$$L\{u(t - 2) \sin (t - 2)\} = e^{-2s}\, L\{\sin t\}$$

$$L\, g(t) = e^{-2s}\, \frac{1}{s^2 + 1^{\,2}}$$

☐ **Example 3-12:** Find the Laplace transform of the function shown in Fig. 3-5.

The time function in Fig. 3-5 has the following equation:

$$f(t) = u(t - 1) \sin (t - 1) + u(t - 2) \sin (t - 2)$$

To find the Laplace transform of this equation, refer to Appendix B and apply operation No. 8. Then refer to Appendix A and employ transforms No. 19 and No. 10. This results in the following Laplace equation:

$$L\, f(t) = \frac{e^{-s}}{s^2 + 1} + \frac{e^{-2s}}{s^2 + 1}$$

☐ **Example 3-13:** Find the Laplace transform of the time function

$$f(t) = tu(t) - (t - 1)\, u(t - 1) - u(t - 3)$$

The above time function has the graphical sum shown in Fig. 3-7. Note, that the sum $tu(t) - (t - 1)\, u(t - 1)$ is the solid line. By adding $-u(t - 3)$ to the solid line, the desired time function, $f(t)$, is obtained.

The Laplace transform of the given time function can be found by referring to Appendix B and applying operation No. 8.

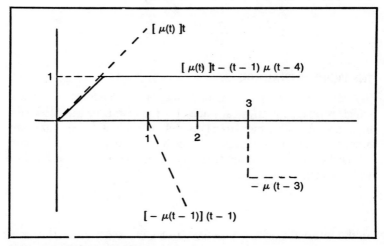

Fig. 3-6. Graph for Example 3-13.

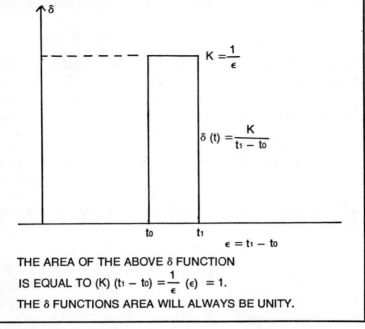

Fig. 3-7. Impulse or delta function.

You must also look in Appendix A and employ transforms No. 19 and No. 29. This forms the following Laplace equation:

$$L\,f(t) = \frac{1}{s^2} - \frac{e^{-s}}{s^2} - \frac{e^{-3s}}{s}$$

THE IMPULSE FUNCTION

Consider the problem of finding the transform of a force which produces a physical impulse—for instance, the blow of a hammer. As a mathematical model of such a force, take any function, $f(t)$, of the time, t, which is zero except in a short interval $t_0 \le t < t_1$. In particular, if K is a constant, the function which equals

$$\frac{K}{t_1 - t_0}$$

in the interval $t_0 \le t < t_1$ is easy to handle and gives satisfactory results for many applications. Such idealized functions are readily expressible in terms of the unit functions and are called *impulse* or *delta* fucntions.

A graph of the impulse function is shown in Fig. 3-7, and the transform of the delta or impulse function is given as transform No. 1 in Appendix A.

THE INITIAL AND FINAL VALUE THEOREMS

The initial value theroem makes it possible for us to calculate the initial value of a time function from the Laplace transform of the function. This calculation may always be accomplished by taking the inverse Laplace transform to obtain the time function and then letting time t approach zero. However, the mechanics of taking the inverse Laplace transform are often tedious, and if the initial value of the time function is the only piece of information desired, the initial value theorem saves the effort of taking the inverse transform.

The *initial value theorem* is stated as follows:

$$\lim_{t \to 0^+} f(t) = f(0^+) = \lim_{s \to \infty} s\,F(s) \qquad \textbf{Equation 3-17}$$

Equation 3-17 exists provided the limit on s F(s) exists. Notice that in the limiting process, t approaches zero from positive values of t, so the initial value of f(t) as defined here is $f(0^+)$, not $f(0^-)$. **Example 3-14:** Find f(t), if the Laplace transform of the time function is

$$F(s) = \frac{as + 1}{s(Ts + 1)}.$$

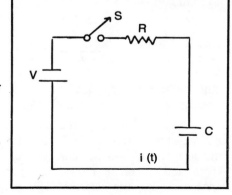

Fig. 3-8. RC series circuit for Example 3-17.

The solve for $f(0^+)$, we must apply Equation 3-17 as follows:

$$f(0^+) = \lim_{s \to \infty} s \frac{as + 1}{s(Ts + 1)} = \lim_{s \to \infty} \frac{as + 1}{Ts + 1}$$

Substitution of $s = \infty$ in this equation results in a fraction ∞/∞. hence, we can apply L'Hospital's rule as follows:

$$f(0^+) = \lim_{s \to \infty} \frac{d(as + 1)/ds}{d(Ts + 1)/ds} = \frac{a}{T}$$

Another theorem which is used in a manner similar to the initial value theorem is one which allows the final value of $f(t)$ to be calculated rapidly from $F(s)$. The theorem is stated as follows:

$$\lim_{t \to \infty} f(t) = f(\infty) = \lim_{s \to 0} s F(s) \qquad \textbf{Equation 3-18}$$

It is important to be aware of the limitations on $F(s)$, because a wrong value for $f(\infty)$ will result if the theorem is applied to a function that is outside those limitations. Stated mathematically, the limitation is that $sF(s)$ must be analytic in the right half s-plane and on the imaginary axis of that plane. This means that $f(t)$ must settle down to a final value if the final value theorem is to be applied; in other words, the system must be stable. More about stability is in Chapter 5.

□ **Example 3-15:** Find the final value, $f(\infty)$, of the time function if the Laplace transform of the time function is the one given in Example 3-14.

To solve for $f(\infty)$, we must apply Equation 3-18 as follows:

$$f(\infty) = \lim_{s \to 0} s \frac{as + 1}{s(Ts + 1)} = \lim \frac{a(0) + 1}{a(0) + 1} = \frac{1}{1} = 1$$

APPLICATION OF LAPLACE TRANSFORMS TO INTEGRODIFFERENTIAL EQUATIONS

After the physical relationships in a linear system are described by means of one or more integrodifferential equations, the analysis of the dynamic behavior of the system may be completed by solving the equations and incorporating the initial conditions into the solution. The examples which follow illustrate the procedure to be used when the Laplace transform is employed to solve linear differential equations with constant coefficients. The Laplace procedure is straightforward and follows the outline stated below:

☐ Describe the system by a set of integrodifferential equations.

☐ Take the Laplace transform of each term in the integrodifferential equation, with the aid of Appendix A and Appendix B. This eliminates t and all the time derivatives and integrals, leaving an algebraic equation in s.

☐ **Example 3-16:** Solve the following differential equation subject to the initial conditions which are stated:

$$\frac{d^3x}{dt^3} \quad + \quad \frac{dx}{dt} \quad = e^{2t} \qquad \qquad \textbf{Equation 3-19}$$

where $x(0^+) = 2$, and $x'(0^+) = x''(0^+) = 0$.

Equation 3-19 can be transformed by employing operation No. 5 in Appendix B, and transform No. 3 in Appendix A as shown.

$$[\,s^3X(s) - s^2x(0^+) - sx'(0^+) - x''(0^+)] + sX(s) - x(0^+) = \frac{1}{s-2}$$

Substitution of the initial values for $x(0^+)$, $x'(0^+)$, and $x''(0^+)$ into the above equation yields the following:

$$s^3X(s) - 2s^2 + sX)(s) - 2 = \frac{1}{s-2}$$

This equation is arranged to yield $X(s)$:

$$X(s) = \frac{2s^3 - 4s^2 + 2s - 3}{s(s^2 + 1)(s - 2)} \qquad \textbf{Equation 3-20}$$

This equation does not have a transform in Appendix A. Therefore, it must be expanded by the method of partial fractions. This results in the following:

$$X(s) = \frac{0.1}{s-2} \quad + \quad \frac{1.5}{s} \quad + \quad \frac{0.4s}{s^2+1} \quad - \quad \frac{0.2}{s^2+1}$$

The solution is obtained by looking in Appendix A for the transforms of the four fractions in the equation. You should be able to apply transform No. 3 to the first fraction, transform No. 18 to second fraction, transform No. 11 to the third fraction, and transform No. 10 to the fourth fraction. The resulting time equation is:

$$x(t) = 0.1e^{2t} + 1.5 + 0.4 \cos t - 0.2 \sin t$$

☐ **Example 3-17:** In the circuit shown in Fig. 3-8, assume that switch S is closed at $t = 0$, and that there is a charge, Q_o, on the capacitor just before the switch is closed. Find the equation for the current, $i(t)$, employing the Laplace transform method.

Write Kirchhoff's voltage law for the circuit and substitute the current-voltage relationship into the voltage equation, as follows:

$$V = v_R + v_C$$
$$= i(t) R + \frac{1}{C} \int i(t) \, dt$$

where $i^{-1}(0) = Q_o$.

Following the same procedure as in Example 3-16, this equation is transformed from time domain t to the Laplace domain s by employing Appendix A and Appendix B. For the integral, refer to Appendix B, operation No. 6. For $i(t)$, substitute $I(s)$ and consider the voltage term, V, as a step function; that is, employ transform No. 18 in Appendix A. Hence, the time equation transforms into the Laplace domain by the following equation:

$$\frac{V}{s} = RI(s) + \frac{1}{C} \frac{I(s)}{s} + \frac{Q_o}{s}$$

Because the time solution for $i(t)$ is desired, solve the previous equation for $I(s)$.

$$I(s) = \frac{(VC - Q_o)/RC}{s + 1/RC}$$

The solution for $i(t)$ is obtained directly from transform No. 3 of Appendix A.

$$i(t) = \frac{(VC - Q_o)}{RC} e^{-t/RC}$$

The Laplace transform method of solving differential equations is a straightforward procedure by which the total solution of a circuit or control system may be found in an orderly fashion. The Laplace method is particularly advantageous when working with control systems, where usually a set of simultaneous equations must be solved. Throughout the remainder of the book, the methods presented in this chapter will be used.

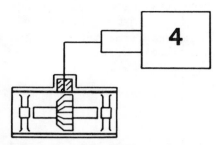

Frequency Response Fundamentals

A control system possesses the property of frequency discrimination, allowing certain frequencies to pass or be transmitted, and other frequencies to be attenuated (rejected). Hence, a control system may be regarded as an *electrical wave filter*, or simply a *filtering device*. The frequency response of a filter is one of the primary questions that must be answered to analyze the performance of a filter. Thus, the same concern of frequency response must be analyzed for a control system. In terms of a control system, one logical question is if the synchro transmitter shaft is driven to oscillate sinusoidally, what will be the behavior of the output of the shaft? It is, of course, very unlikely that the input to the system will be sinusoidal in practice, but by working with frequency response, we shall show that interpretation may be made also on the transient behavior of the system.

In this chapter, the basic ideas of filters employing the *operational approach to circuit analysis* will be presented in order to explain an easy method of obtainining the frequency response of filters and eventually control systems.

DEFINITIONS AND CONCEPTS OF FILTERS

Three common types of filters are the low-pass, high-pass and bandpass filters. A low-pass filter allows signals, which can be called $F(j\omega)$, to be transmitted up to a certain maximum frequency. Remember that ω (omega) $= 2\pi f$. Beyond the cutoff frequency defined by ω_c, the signal $F(j\omega)$ is rejected or attenuated. Treble controls and scratch filters are examples of

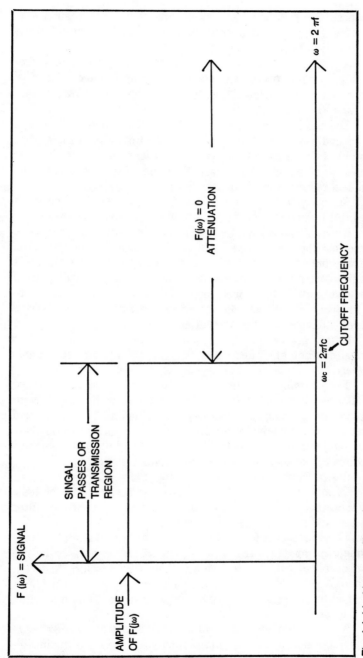

Fig. 4-1. Ideal low-pass filter response.

69

low-pass filter circuits. An *ideal* low-pass filter response is shown in Fig. 4-1, and an approximate nonideal low-pass filter is illustrated in Fig. 4-2.

A high-pass filter blocks frequencies below its cutoff frequency, f_c, while transmitting desired signals above f_c. The ideal and nonideal high-pass filter characteristics are shown in Fig. 4-3. and Fig. 4-4, respectively.

The bandpass filter selects a range of frequencies between upper and lower cutoff frequencies. The tuning on a radio is an example of a variable bandpass filter. Figure 4-5 illustrates the response of an ideal bandpass filter and Fig. 4-6 shows the nonideal response of a bandpass filter.

An *asymptote* is a line on the frequency response curve of the filter that approximates the actual frequency response.

Break frequency is the point on the frequency response curve that the filter drops 3 dB, or 0.707, of its largest value on the way out of the passband. The preceding definition considers the filter frequency response to roll off to a smaller value out of the passband. It is entirely possible for a break frequency to occur when the filter will experience a rise in the response curve. This is also called the cutoff frequency.

Filters are designed by placing RC, RL, and RLC sections one after another. These sections are referred to as *cascade sections*.

The *damping* of a filter is an index of its tendency toward oscillation. Practical damping values range from 2 to 0, with zero damping being the value for an oscillation. Highly damped filter sections combine to produce smooth filters with good overshoot and transient response. Slightly damped filters combine to produce filters with sharp rejection characteristics.

A *decade* is a 10-to-1 frequency interval.

Decibels are a logarithmic way of measuring gain or loss. Decibels are defined as 20 $\log_{10}$ of a voltage ratio. In filter design, dBs refer only to a voltage ratio.

The expression for *gain* refers to the ratio of output voltage divided by input voltage. Often, gain will be expressed by the symbol, $N(j\omega)$, which refers to a straight voltage ratio. Also, the symbol N_{2b} may be used, which means the voltage ratio has been changed to decibels. The symbol $N(j\omega)$ is read "N as a function of j omega."

A 2-to-1 frequency interval is called an *octave*. *Omega* (ω) is the symbol that defines angular velocity associated with sinusoidal functions. Remember, $\omega = 2\pi f$.

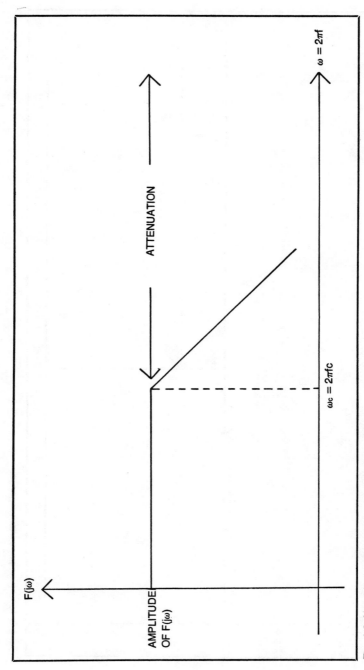

Fig. 4-2. Non-ideal approximation of a low-pass filter.

71

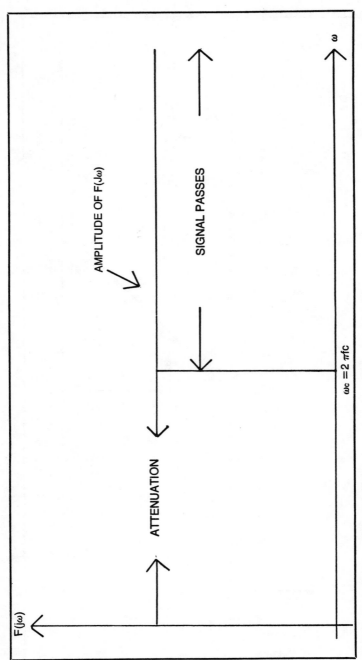

Fig. 4-3. Ideal high-pass filter response.

72

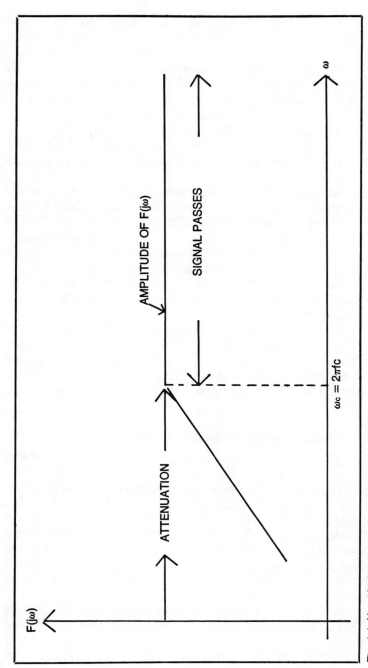

Fig. 4-4. Non-ideal approximation of a high-pass filter.

73

A *normalized* filter is one wnose component values and frequency response curve are adjusted to a convenient frequency.

The *order* of a filter governs the strength of its fall off with frequency. For example, a third-order low-pass filter falls off as the *cube* of frequency, which means at 3 times 6dB/octave = 18 dB/octave, or 3 times 20 dB/decade = 60 dB/decade.

Q is the inverse of the damping value. In other words $Q = 1/\zeta$. The value of Q is used to measure the bandwidth of a second-order bandpass section. Practical Q values range from less than one to several hundred.

Scaling is denormalizing a filter by changing its frequency or component values. Impedance level scaling is increased by multiplying all resistors and dividing all capacitors by a desired factor. To double the frequency, decrease all capacitor values by 2, or cut resistor values by 2. If you decrease both the resistor and capacitor values, the frequency will be quadrupled.

The *sensitivity* of a filter is a measure of how accurate the component tolerances must be to get a response within certain limits of what is desired.

The *transfer function* of a filter is simply what the output of a filter is compared to the input. The transfer function is usually expressed as the ratio of V_{out}/V_{in}. The transfer function could include both the amplitude and phase information. The phase information refers to the phase shift from the input to the output.

OPERATIONAL APPROACH TO CONTROL SYSTEMS

The operational approach of circuit analysis, which includes filters, requires little past knowledge of AC or DC circuit methods and requires some knowledge of algebra. The operational approach to dealing with resistors, inductors, and capacitors is to define the transformed impedance of R, L, and C, which is respectively R, sL, and 1/sC. The transformed admittances of the components is the reciprocal of the transformed impedances, which are 1/R, 1/sL, and sC. Figure 4-7 shows the schematic representation of R, L, and C and their transformed impedance.

The quantity, s, is called the *complex variable*. The quantity s has a real part, σ (sigma), and an imaginary part, $j\omega$. In filter design, the real part simna is usually omitted, and $j\omega$ is used **in** place of the complex variable, s.

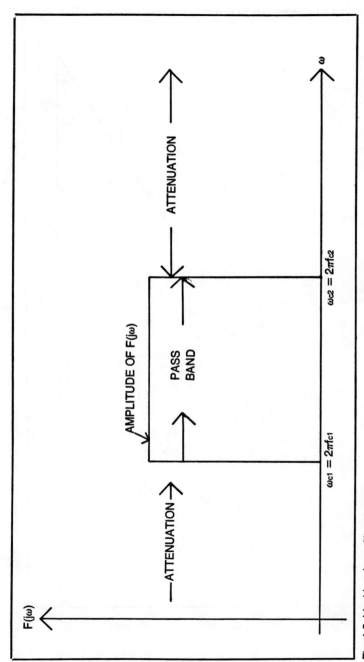

Fig. 4-5. Ideal bandpass filter response.

75

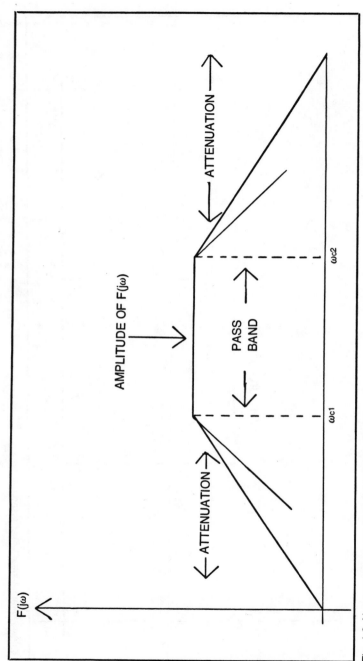

Fig. 4-6. Non-ideal approximation of a high-pass filter.

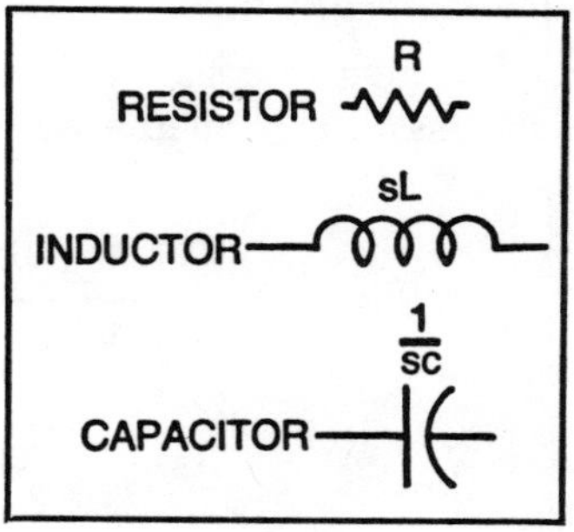

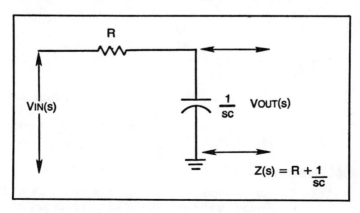

Fig. 4-7. Transform impedance of a resistor, inductor, and capacitor.

RESISTOR R

INDUCTOR sL

CAPACITOR $\dfrac{1}{sC}$

The circuit in Fig. 4-8 is a simple RC low-pass filter circuit of the first order (notice it only has one capacitor). The transformed impedance, Z(s), of the circuit is found by adding together the transformed impedance of R and the transformed impedance of the C, since the resistor and capacitor are in series with each other. Hence, the equation for the transformed impedance of circuit in Fig. 4-8 is:

$$Z(s) = R + \frac{1}{sC}$$

Examples of finding transformed impedances of other circuits are shown in Fig. 4-9. The simple series and parallel circuit principles of finding equivalent impedances are carried over into finding transformed impedance as illustrated in Fig. 4-9.

$$V_{IN}(s) \qquad \frac{1}{sC} \qquad V_{OUT}(s)$$

$$Z(s) = R + \frac{1}{sC}$$

Fig. 4-8. Low-pass filter circuit.

TRANSFER FUNCTION

Looking at Fig. 4-8, you can write an expression for the high-pass filter, called a transfer function, T(s), and defined as follows:

$$T(s) = \frac{V_{out}(s)}{V_{in}(s)}$$

$$V_{out}(s) = I(s)\ \frac{1}{sC}$$

$$V_{in}(s) = I(s)\ Z(s) = I(s)\left[R + \frac{1}{sC} \right]$$

$$T(s) = \frac{I(s)\ \dfrac{1}{sC}}{I(s)\left[R + \dfrac{1}{sC} \right]}$$

$$T(s) = \frac{\dfrac{1}{sC}}{\dfrac{sRC + 1}{sC}}$$

$$T(s) = \frac{1}{1 + sRC}$$

$$T(s) = \frac{\dfrac{1}{RC}}{\dfrac{1}{RC} + \dfrac{sRC}{RC}}$$

$$T(s) = \frac{\dfrac{1}{RC}}{s + \dfrac{1}{RC}} \qquad \textbf{Equation 4-1}$$

Equation 4-1 represents the transfer function of the high-pass filter circuit in Fig. 4-5.

BODE DIAGRAMS

Once the transfer function for a filter is determined, the amplitude and phase characteristic of the filter is next plotted. A very useful technique for obtaining a close approximation of the

78

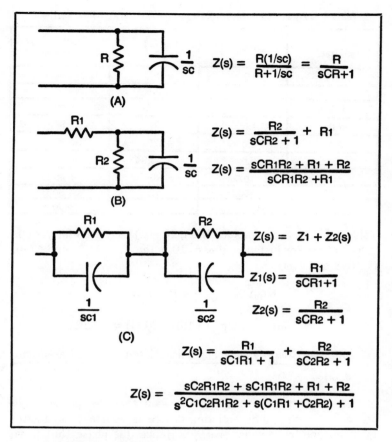

Fig. 4-9. Other transformed circuits.

amplitude and phase curve from the transfer function is called the Bode diagram. Consider as a first example the transfer function for a pure differentiating device.

$$G(s) = s \qquad \text{Equation 4-2}$$

The frequency response of this device is obtained by substituting $j\omega$ for s in the transfer function. The amplitude and phase shift are as follows:

$$N(\omega) = |G(j\omega)| = \omega \qquad \text{Equation 4-3a}$$

$$\phi(\omega) = |G(j\omega)| = +90° \qquad \text{Equation 4-3b}$$

79

Table 4-1. Frequency Response Data for the Transfer Function, G(s) = s.

ω rads/sec.	N(ω)	N_{dB} (ω) in dB	φ (ω) in degrees
0.0	0.0	− ∞	+ 90
0.1	0.1	− 20	+ 90
0.5	0.5	− 6.02	+ 90
1.0	1.0	0	+ 90
2.0	2.0	+ 6.02	+ 90
5.0	5.0	+ 13.94	+ 90
10.0	10.0	+ 20	+ 90
100.0	100.0	+ 40	+ 90
∞	∞	∞	+ 90

Table 4-1 lists the values for $N(\omega)$, N_{dB} (ω). and $\phi(\omega)$ for various values of ω. For instance, if $\omega = 10$ in the Table 4-1, find $N(\omega)$, N_{dB} (ω), and $\phi(\omega)$.

$$N(\omega) = |G(j\omega)| = |10| = + 10$$

$$N_{dB}\ (\omega) = 20\ \log_{10}\ (+10) = 20\ (1) = 20\ dB$$

$$\phi(\omega) = \underline{/G(j\omega)} = \underline{/\text{phase angle associated with a differentiator}}$$
$$\phi(\omega) = + 90°$$

At this time, it can only be said that the phase angle associated with a differentiating device is constant for all frequencies and has a value of +90°. Hence, the phase shift curve for a differentiating device is simply a flat line curve for all input frequencies, and its value is +90°.

The results in Table 4-1 is plotted on semi-log graph paper. The quantity N_{dB} (ω) plotted against the frequency variable ω is a straight line, as illustrated in Fig. 4-10. The slope of the straight line is measured as the change in dB level between two frequency points that are separated by a given distance along the horizontal axis. This distance cannot be measured in radians per second because a given distance at the low frequency end of the scale will represent a change of very few radians per second, while the same distance at the high-frequency end will represent a change of very many radians per second. However, a given distance along the horizontal axis will represent the same percentage change in frequency, no matter where that distance is taken along the frequency scale. Therefore, the slope of the N_{dB} (ω) curve is expressed in dB per percentage change in frequency. A change in frequency of 100 percent from, say $\omega = 1$ to $\omega = 2$, is called an *octave,* so for this curve, the amplitude N_{dB} (ω) changes at a rate of 6.02 dB/octave. However, for a very close approximation, we refer to a 6 dB/octave change.

A change in frequency by a factor of 10 (from $\omega = 1$ to $\omega = 10$) is called a decade, so in this system, the curve N_{dB} (ω) has a slope of 20 dB/decade. Remember that 6 dB/octave is the same slope as 20 dB/decade.

The plot of N_{dB} (ω) is called a Bode plot after the person who developed the diagram. Let us analyze the Bode plot and phase diagram of the function shown below;

$$G(s) = Ks \qquad \textbf{Equation 4-4}$$

The above equation is the transfer function of the differentiating device, multiplied by a constant, K. The amplitude and phase shift are as follows:

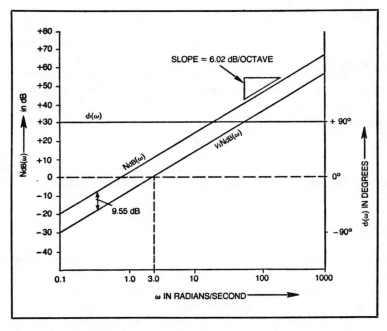

Fig. 4-10. Bode plot of $G_1(j\omega) = j\omega$ and $G_2(j\omega) = \frac{1}{3}\ j\omega$.

$$N(\omega) \quad = G(j^\omega) \quad = k\omega \qquad \qquad \text{Equation 4-5a}$$

$$\phi(\omega) \quad = G(j^\omega) \quad = +90° \qquad \qquad \text{Equation 4-5b}$$

A table of frequency response data for this system would be the same as that shown in Table 4-1, except that the numbers in the $N(\omega)$ column would be K times those shown, and the numbers in the $N_{dB}(\omega)$ column would be those in Table 4-1, plus $20 \log_{10}$ K. From this, it can be concluded that if a transfer function is multiplied by a constant, the phase shift of the system is *unaffected* but the amplitude ratio is increased or decreased by the constant factor. On the Bode diagram, this increase or decrease shows up in the amplitude curve as a vertical shift by the number of decibels corresponding to the constant multiplier. For example, if K = ⅓, the amplitude curve would be shifted downward 9.55 dB, since $20 \log_{10}$ (⅓) = -9.55 dB. The Bode plot of this system is shown on the graph in Fig. 4-10. However, note the slope of the response has not changed from the first example when K = 1, to the second example when K = ⅓.

As another example of the straight line Bode plot, consider the following transfer function:

$$G(s) = \frac{K}{s} \qquad \qquad \text{Equation 4-6}$$

The above equation is the transfer function of an integrator device. The amplitude and phase shift of the integrator are found by substituting jω for s in the transfer function. Hence, the transfer function becomes:

$$G(j\omega) = \frac{K}{j\omega}$$

Next, we find the amplitude and phase shift of this equation.

$$N(\omega) = |G(j\omega)| = \frac{K}{\omega} \qquad \qquad \text{Equation 4-7a}$$

$$\phi(\omega) = G(j\omega) = -90° \qquad \qquad \text{Equation 4-7b}$$

The amplitude ratio is inversely proportional to ω, so that at ω = K, then $N(\omega) = 1$, which is equal to 0 dB. The slope of the $N_{dB}(\omega)$ curve will be -6.02 dB/octave or -20 dB/decade, as shown on the curve of Fig. 4-1. The Bode diagram has been drawn for K = 4. At this time we can only say that the phase angle associated with the integrator device is constant for all frequencies and has a value of $-90°$, which is plotted on the graph of Fig. 4-11.

Find the frequency response (amplitude response) and phase shift of a double integrator. The transfer function or a double integrator is given in Equation 4-8.

$$G(s) = \frac{K}{s^2} \qquad \textbf{Equation 4-8}$$

Substitute $s = j\omega$ in Equation 4-8.

$$G(j^\omega) = \frac{K}{(j\omega)^2} = \frac{K}{j^2 \omega^2} = \frac{K}{=\omega^2}$$

For this transfer function, the frequency response and phase shift are given by the following equations:

$$N(\omega) = |G(j\omega)| = \frac{K}{\omega^2} \qquad \textbf{Equation 4-9a}$$

$$\phi(\omega) = \underline{/G(j\omega)} = -180° \qquad \textbf{Equation 4-9b}$$

Again, the Bode plot for $N_{dB}(\omega)$ is a straight line having a negative slope, since $N(\omega)$ decreases with an increase in ω but the slope will be -40 dB/decade, or -12 dB/octave, since an

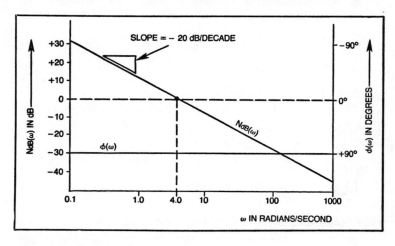

Fig. 4-11. Bode plot of $G(j\omega) = 4/j\omega$.

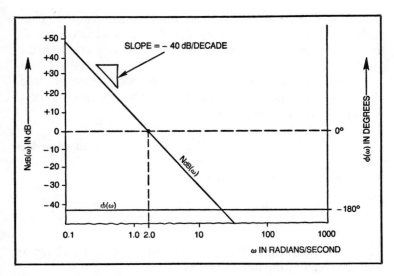

Fig. 4-12. Bode plot of $G(j\omega) = 4/(j\omega)^2$.

increase of one decade in ω causes $N(\omega)$ to change by a factor of 100. The amplitude and phase curves are drawn for $K = 4$ in Fig. 4-12. At this time, it can only be said that the phase angle associated with the integrator device is constant for all frequencies and has a value of $-180°$, which is plotted on the graph of Fig. 4-12.

A comparison of the curves in Figs. 4-10 through 4-12 indicates that the Bode plot for a transfer function of the form:

$$G(s) = K \, s^n \qquad \textbf{Equation 4-10}$$

In this equation, the value of n can be positive or negative. The quantity n has the following properties:

☐ $N_{dB}(\omega)$ will be a straight line having a slope of 20n dB/decade. Of course, if n is negative, the slope of $N_{dB}(\omega)$ will be negative.

☐ $N_{dB}(\omega)$ will cross the zero dB line at the frequency $\omega = K^{-1/n}$.

☐ $\phi(\omega)$ will be constant at 90n°.

BODE DIAGRAMS OF FIRST-ORDER TRANSFER FUNCTIONS

Next, consider transfer functions having Bode plots which are not straight lines, but which may be approximated by straight lines. We will accomplish this by considering the *asymptotic behavior* of transfer functions of the first order at extremely low and extremely high freqeuncies.

Consider the transfer function with the first-order form shown.

$$G(s) = K(s + a) \qquad \textbf{Equation 4-11}$$

The order of the transfer function is determined by the highest power of s. In Equation 4-11, quantity s has the power of 1; hence, there is a first-order transfer function. If the highest power of s is 2, there would be a second-order transfer function, etc.

To obtain the frequency (amplitude) response and phase characteristics, let s = jω and substitute this value for s in the first-order transfer function of Equation 4-11. The result is shown below:

$$G(j\omega) = K(j\omega + a) \qquad \textbf{Equation 4-12}$$

Next, find the absolute value (magnitude) of Equation 4-12.

$$N(\omega) = |G(j\omega)| = K(\omega^2 + a^2)^{\frac{1}{2}} \qquad \textbf{Equation 4-13a}$$

The absolute value or magnitude of Equation 4-12 is a vector since quantity *a* represents a real number, and the quantity jω is a complex number situated at an angle of 90° from *a*. Figure 4-13 illustrates how the magnitude of the first-order transfer function is obtained.

The phase shift of the first-order transfer function is also illustrated in Fig. 4-13. The phase shift angle has a tangent which is equal to ω/a. In other words, tan ϕ equals the side

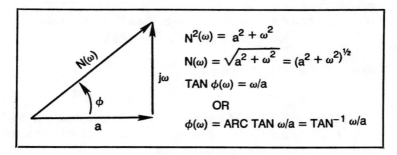

Fig. 4-13. Vector diagram of N(ω) for the transfer function G(s) = s + a.

opposite the angle ϕ in Fig. 4-13, divided by the side adjacent the angle ϕ in Fig. 4-13. In equation form, tan ϕ = ω/a. To express angle $\phi(\omega)$, the inverse tangent function must be found, which is called the arc tan or written $\tan^{-1}$. This function can be expressed:

$$\phi(\omega) = \underline{/G(j\omega)} = \tan^{-1} \frac{\omega}{a} = \text{arc tan} \frac{\omega}{a} \qquad \textbf{Equation 4-13b}$$

Consider how the behavior of $N(\omega)$ can be graphed. First consider when the frequencies are very low; that is, when ω is much smaller than a. At these low frequencies, $N(\omega)$ will be approximately equal to the product of K times a. In equation form, this approximation can be written as:

$$N(\omega) \cong Ka \text{ when } \omega << a$$

This equation is a straight line of zero slope with a dB value of $20 \log_{10}$ (Ka), as shown in the Bode diagram of Fig. 4-14.

Next, consider when the frequencies are very high; that is, when ω is much larger than a. At these high frequencies, $N(\omega)$ will be approximately equal to the product of K times ω. In equation form, this approximation can be written as:

$$N(\omega) \cong K\omega \text{ when } \omega >> a$$

This equation is a straight line having a slope of 20 dB/decade and passing through the $20 \log_{10}$ (Ka) point at ω = a, as illustrated in Fig. 4-14. Thus, the two straight lines, shown as dotted lines in Fig. 4-14 that intersect at ω = a, comprise a very close approximation to the amplitude response curve, $N_{dB}(\omega)$, which is shown as a solid curve in Fig. 4-14.

As already shown, the approximation of the overall frequency response is very good at very low and at very high frequencies, but now investigate the error at frequencies in the neighborhood of ω = a. In Fig. 4-14, this occurs at the intersection of the two dotted lines. At ω = a, the following results:

$$N(\omega)\big|_{\omega} = a = K(a^2 + a^2)^{\frac{1}{2}} = \sqrt{2} \, Ka$$

In terms of dB we have the value;

$$N_{dB}(\omega)\big|_{\omega} = a = 20 \log_{10}(Ka) + 20 \log_{10} \sqrt{2}$$

Finding the value of the second term in this equation will indicate how many dB the actual curve is above the intersection of the two straight line approximations.

$$20 \log_{10} \sqrt{2} = 3.01 \text{ dB}$$

Hence, the Bode plot of the frequency response actually passes through a point that is 3.01 dB above the intersection of the two straight lines. The quantity $\omega_c = a$ is called the *break frequency* or the *corner frequency* of this transfer function.

At a frequency one octave below the break frequency—that is, at $\omega = a/2$—the amplitude ratio is:

$$N(\omega)/_{\omega = a/2} = K\left[\frac{a^2}{4} + a^2\right]^{\frac{1}{4}} = (\sqrt{5/4})\,Ka$$

which, expressed in decibels, is

$$N_{dB}(\omega)|_{\omega = a/2} = 20\log_{10}(Ka) + 20\log_{10}\sqrt{5/4}$$

But the value of $20\log_{10}\sqrt{5/4} = 0.969$. Hence, the Bode plot passes through a point which lies 0.969 dB above the straight line asymptote at $\omega = a/2$.

Similarly, at a frequency one octave above the break frequency at $\omega = 2a$, the Bode plot will pass through a point which is 6.99 dB above the low-frequency asymptote or 0.97 dB above

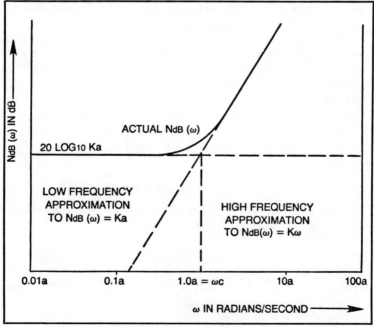

Fig. 4-14. Straight line approximations to $N_{dB}(\omega)$ on the bode chart for the transfer function, $G(s) = K(s + a)$.

the high-frequency asymptote. When the asymptotes are drawn on the graph as shown in Fig. 4-14, and the three points, $\omega_1 = a/2$, $\omega_2 = a$, and $\omega_3 = 2a$, are spotted on the graph, it is possible with the aid of a French curve to draw the graph for the actual function $N_{dB}(\omega)$. The actual graph of $N_{dB}(\omega)$ is shown by the solid line curve in Fig. 4-14. In many problems it is not necessary to have an extremely accurate graph, and the deviations of the straight line asymptotes from the actual $N_{dB}(\omega)$ curve may be taken as 1, 3, and 1 dB, instead of the more accurate 0.969, 3.01, and 0.97 dB at the three frequencies just considered.

A plot of the phase shift curve is shown in Fig. 4-15. A plot of the phase shift curve given by Equation 4-16 is an arc tangent curve, as shown in Fig. 4-15. At the break frequency, the phase shift is exactly 45 degrees. At $\omega = a/2$, the phase shift is about 26.57 degrees, and at $\omega = 2a$, the phase shift is about 63.43 degrees. A straight line may be drawn through these three points on the phase shift curve of Fig. 4-15. The striaght line, which is shown broken, will intersect the $\phi(\omega) = 0$ line (which may be considered to be the low-frequency asymptote) at $\omega = 0.184a$, and it will intersect the $\phi(\omega) = 90$-degree line (the high-frequency asymptote) at $\omega = 5.43a$. These two frequencies, 0.184a and 5.43a, can be considered as break frequencies on the $\phi(\omega)$ curve. A simple calculation will show that the phase shifts at these two frequencies are 10.435 degrees and 79.565 degrees. These figures form a basis for making a straight line approximation to the phase shift curve. Using Equation 4-13b, other straight line approximations to the phase shift curve could be derived. For example, the tangent to the $\phi(\omega)$ curve at $\omega = a$ could be used to approximate the midfrequency phase shift curve.

POLE AND ZERO VALUES FOR THE TRANSFER FUNCTION

As a further example of the straight line approximation technique, consider the transfer function:

$$G(s) = \frac{K(s + a)}{s + b} \qquad \textbf{Equation 4-14a}$$

The transfer function defined by Equation 4-14a can be decomposed as the product of three transfer functions as follows:

$$G(s) = \left[\frac{Ka}{b}\right] \left[\frac{s + a}{a}\right] \left[\frac{b}{s + b}\right] \qquad \textbf{Equation 4-14b}$$

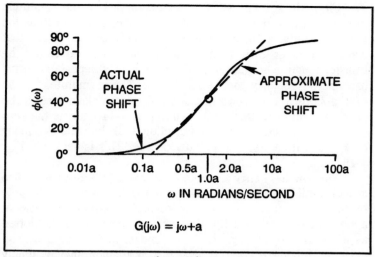

Fig. 4-15. Phase shift curve for $G(j\omega) = j\omega + a$.

The *poles* of the above transfer function are found by equating the denominator of Equation 4-14b to zero and solving for s as follows:

$$s + b = 0$$

and

$$s = -b, \text{ the pole of Equation 4-14b}$$

The *zeros* of the transfer function in Equation 4-14b are found by equating the numerator of Equation 4-14b to zero and solving for s as follows:

$$s + a = 0$$

and

$$s = -a, \text{ a zero of Equation 4-14b}$$

If a transfer function is composed of *several first-order poles and zeros*, the composite frequency response curve can be drawn by considering the individual transfer functions containing the poles and zeros, drawing their frequency response curves as discussed in the preceding sections, and combining their results to obtain the overall frequency response. Every pole and zero defines a break frequency, which is numerically equal to the pole or zero value. The task at hand is to then draw the frequency response curve for the transfer function of Equation 4-14b.

First, draw a straight line approximation to the amplitude ratio for each of the three simple transfer functions in Equation 4-14b. See Fig. 4-16. Assume that a is greater than b, and that Ka/b is greater than 1. Also, it is convenient to factor the transfer function so that each of the frequency-dependent terms has a zero frequency gain of unit (0 dB). The asymptotes for the $(s + a)/a$ term are drawn in exactly the same manner as in the previous sections, which is shown in Fig. 4-14. Hence, the zero of a transfer function will break from the 0 dB line at a frequency of $\omega = a$, with a slope of +20 dB/decade. This is shown in Fig. 4-13.

The low-frequency asymptote for the $b/(s + b)$ term is also at 0 dB until break frequency $\omega = b$ is reached, which is the pole of the transfer function. The low-frequency asymptote breaks at a slope of -20 dB. Finally, the constant term, Ka/b, is plotted as a constant of $20 \log_{10}$ Ka/b, as shown in the graph of Fig. 4-16. A composite curve is drawn in a heavier line by simply adding the three component approximations. The actual magnitude curve will lie close to this composite straight line approximation, as indicated by the dotted lines in Fig. 4-16.

In summary, if a transfer function is composed of poles and zeros, the composite frequency response curve can be drawn by adding the response curves of each pole term, zero term, and constant term of the original transfer function. One straight line approximation is required for each of the individual components at the break frequency corresponding to that component. The break frequency is numerically equal to the pole or zero value, and the high-frequency asymptote breaks *upward* for the component contributed by a zero of the transfer function, and it breaks *downward* for the component contributed by a pole of the transfer function.

The phase shift curves for the individual transfer functions are shown in Fig. 4-17. Notice that the phase shift for the $b/(s + b)$ term is negative or *lagging*, and the phase shift for the $(s + a)/a$ term is positive or *leading*. The phase shift for the constant term is zero, of course. The composite phase shift curve is formed by adding the three individual curves in the same manner as the composite amplitude curve was formed in Fig. 4-16.

ANOTHER EXAMPLE OF PLOTTING
APPROXIMATE AMPLITUDE RESPONSE

As a final example of approximating the amplitude characteristic, consider the transfer function:

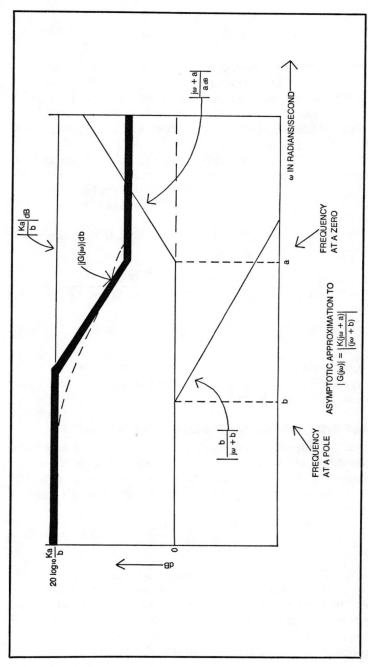

Fig. 4-16. Asymptotic approximation to $|G(j\omega)| = |K(j\omega + a/(j\omega + b)|$.

91

$$G(s) = \frac{40(s + 11)(s + 24)}{(s + 2)(s + 6)(s + 40)}$$ **Equation 4-15**

The zeros of the transfer function in Equation 4-15 occur at s = -1 and s = -24. Remember that the zeros of the transfer function define break frequencies at $\omega = 1$ and $\omega = 24$, as shown in Fig. 4-18. The straight line approximation of the zero break frequency break upward at a $+20$ dB/decade slope. The poles of the transfer function in Equation 4-15 occur at s = -2, s = -6, and s = -40. Remember that the poles of the transfer function define break frequencies at $\omega = 2$, $\omega = 6$, and $\omega = 40$, as shown on the graph of Fig. 4-18. The straight line approximation of the pole break frequencies break down at a -20 dB/decade slope.

The initial value of 6 dB was found through the arrangement of the transfer function as follows:

$$G(s) = \frac{(40)(1)(24)}{(2)(6)(40)} \frac{(s+1)}{1} \frac{(s+24)}{24} \frac{2}{(s+2)} \frac{6}{(s+6)} \frac{40}{(s+40)}$$

Then the value of $20 \log_{10} \frac{(40)(1)(24)}{-(2)(6)(40)}$

equals 6 dB is the beginning of the amplitude response of the curve in Fig. 4-18. The dotted line shows approximately where the actual magnitude curve would lie.

FREQUENCY AND IMPEDANCE SCALING OR NORMALIZING

The frequency response of a given active filter can be *shifted, scaled,* or *normalized* to a different region of the frequency axis by dividing *either* the resistor or capacitor values by a frequency normalizing factor, *u*. The quantity *u* is expressed as follows:

$$u = \frac{\text{Desired frequency}}{\text{Existing frequency}}$$ **Equation 4-16**

Both the numerator and denominator must be expressed in the same units. In other words, either frequency (f) must be used in both numerator and denominator or frequency (ω) must be used in both numerator and denominator. Usually, 3 dB points (break frequency locations) are selected as reference frequency for low-pass or high-pass filters, and the center frequency is

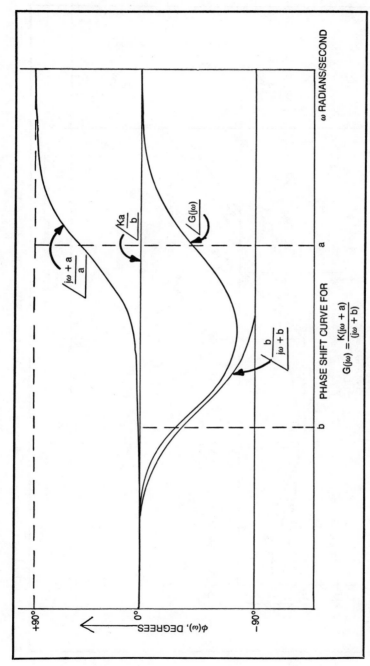

Fig. 4-17. Phase shift curve for $G(j\omega) = K(j\omega + a)/(j\omega + b)$.

93

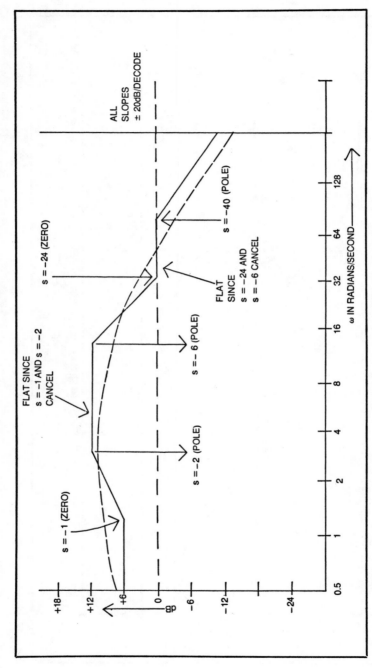

Fig. 4-18. Amplitude ratio curve (approximation) for G(s) = 40(s + 1)(s + 24)/(s + 2)(s + 6)(s + 40).

selected for a bandpass filter. The following example will illustrate the use of the frequency normalizing factor, u.

□ **Example 4-1:** For the normalized low-pass filter and associated frequency response shown in Fig. 4-19, find the denormalized low-pass filter circuit and the associated frequency response if the break freqeuncy occurs at 1000 Hz.

First, calculate the frequency normalizing factor, u.

$$u = \frac{2\pi \ (1000) \ \text{rads/second}}{1 \ \text{rad/second}} = 6280$$

Next, divide the capacitor values by quantity u, which results in the denormalized filter circuit of Fig. 4-20A, and the response curve shown in Fig. 4-20B.

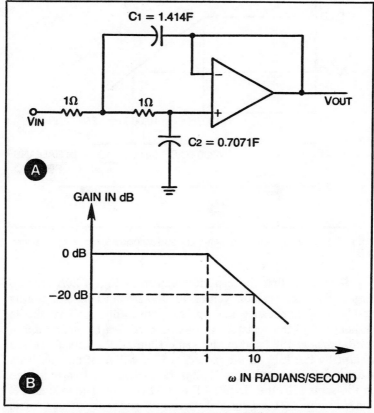

Fig. 4-19. Normalized low-pass filter at A and associated frequency response at B.

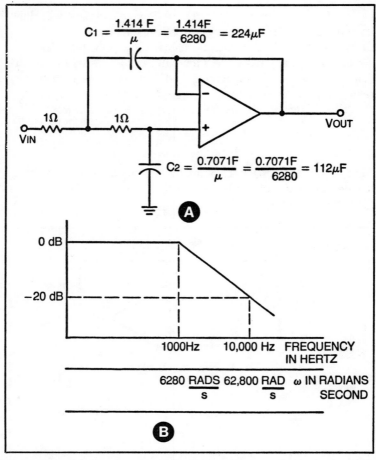

Fig. 4-20. Denormalized low-pass filter at A and associated frequency response at B.

It is clear that the capacitor values in Fig. 4-20A are too large, while the resistor values are too small. This situation is resolved by impedance scaling. An active filter will retain its response if all the resistor values are *multiplied* by an *impedance scaling factor*, ISF, and at the same time all the capacitors are *divided* by the ISF. In the circuit of Fig. 4-20A, if the resistors are multiplied by an ISF = 10,000 and the capacitors are divided by the same ISF, the circuit of Fig. 4-21 results. The values of R and C in Fig. 4-21 are certainly practical, and these values retain the same frequency response as shown in Fig. 4-20B.

The process of frequency and impedance scaling can be combined into the following equations:

$$C_p = \frac{C_n}{(\mu)(\text{ISF})}$$ **Equation 4-17**

C_p = practical capacitor value

C_n = normalized capacitor value

μ = frequency normalizing factor

ISF = impedance scaling factor

$$R_p = R_n\,(\text{ISF})$$

R_p = practical resistor value

R_n = normalized resistor value   **Equation 4-18**

SECOND-ORDER TRANSFER FUNCTION AMPLITUDE AND PHASE RESPONSES

If a transfer function should contain a term with s^2, the transfer function cannot be decomposed to a product or individual first-order transfer functions. Therefore, a straight line approximation to the frequency response characteristics will not

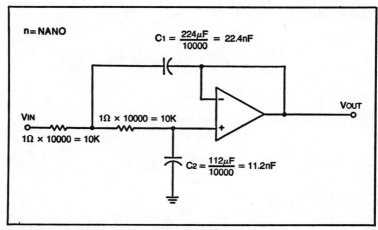

Fig. 4-21. Denormalized low-pass filter with impedance scaling.

be quite as simple as it is when only first-order transfer functions are present.

Consider the second-order transfer function:

$$G(s) = \frac{\mu_n{}^2}{s^2 + 2\zeta\omega_n s + \omega_n{}^2} \qquad \text{Equation 4-19}$$

Both the amplitude response, $N(\omega)$, and the phase shift, $\phi(\omega)$, depend upon two quantities in equation 4-19. The first quantity is called the *damping*, d. The second quantity is called the *natural* frequency, ω_n. It is convenient to normalize the amplitude response, $N(\omega)$ and the phase shift, $\phi(\omega)$, with respect to ω_n. We start with substituting $s = j\omega$ in equation 4-19.

$$G(j\omega) = \frac{\omega_n{}^2}{(j\omega)^2 + 2\zeta\omega_n(j\omega) + \omega_n{}^2}$$

$$G(j\omega) = \frac{\omega_n{}^2}{(-\omega^2) + j\,2\zeta\omega_n\omega + \omega_n{}^2}$$

Divide numerator and denominator by $\omega_n{}^2$

$$G\left(\frac{j\omega}{\omega_n}\right) = \frac{1}{-\left(\dfrac{\omega}{\omega_n}\right)^2 + j2\zeta\left(\dfrac{\omega_n}{\omega_n}\right)\left(\dfrac{\omega}{\omega_n}\right) + \left(\dfrac{\omega_n}{\omega_n}\right)^2}$$

Normalizing the frequency by letting $\mu = \dfrac{\omega}{\omega_n}$,

$$G(j\mu) = \frac{1}{-(\mu)^2 + j2\zeta(1)(\mu) + 1^2}$$

$$= \frac{1}{(1 - \mu^2) + j2\zeta\mu}$$

Find the magnitude of this equation:

$$|G(j\mu)| = N(\mu) = \frac{1}{\left[(1 - \mu^2)^2 + (2\zeta\mu)^2\right]^{1/2}} \qquad \text{Equation 4-20}$$

Next, find the phase shift:

$$\phi(\mu) = -\tan^{-1}\left(\frac{2\zeta\mu}{1 - \mu^2}\right) \qquad \text{Equation 4-21}$$

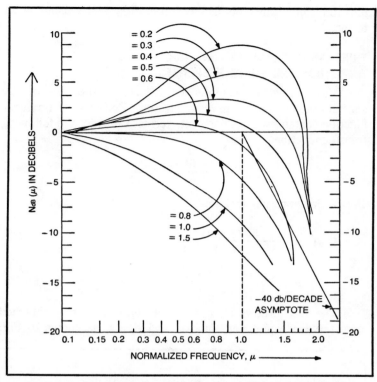

Fig. 4-22. Bode plot of amplitude ratio of second-order transfer function.

The amplitude and phase shift depend upon two variables, ζ and μ. The amplitude response, $N(\mu)$, and the phase response, $\phi(\mu)$, are accurately plotted for various values of ζ on Bode charts in Fig. 4-22 and Fig. 4-23, respectively. The high-frequency asymptote for all curves in Fig. 4-22 is a straight line passing through the 0 dB line at $\mu = 1$ with a slope of -40dB/decade. Notice also that as the damping ζ is decreased, the peaking of the response curve around $\mu = 1$ becomes flatter, but with a sacrifice of getting into the attenuation band more slowly. If the damping were equal to zero ($\zeta = 0$), an infinite peak response or oscillation would occur.

When the normalized frequency response curves for the second-order transfer function are available, it is possible to make an asymptotic approximation to the frequency response characteristics of any transfer function which is the ratio of two algebraic polynomials in s. Note that this type of transfer function can always be factored into first and second order terms.

THIRD-ORDER TRANSFER FUNCTION
AMPLITUDE AND PHASE RESPONSES

The following example will illustrate the use of the second-order response curves in making a Bode plot of a transfer function containing s^3 as one of its terms. Consider the following transfer function:

$$G(s) = \frac{50 \, (s + 4)}{s^3 + 4s^2 + 100 \, s}$$

Factor this equation as follows:

$$G(s) = \frac{50 \, (s + 4)}{s(s^2 + 4s + 100)}$$

$$= \frac{(50)(4)}{(100)} \quad \frac{(s + 4)}{(4)} \quad \frac{(1)}{(s)} \quad \frac{(100)}{(s^2 + 4s + 100)}$$

The first three terms in the above equation can be sketched onto the Bode plot very quickly and easily, as shown in Fig.

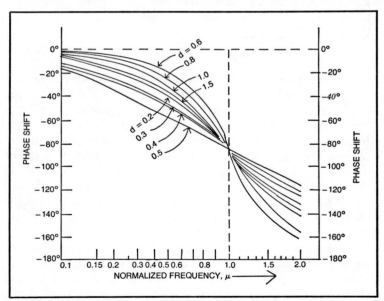

Fig. 4-23. Bode plot of phase shift from a second-order transfer function.

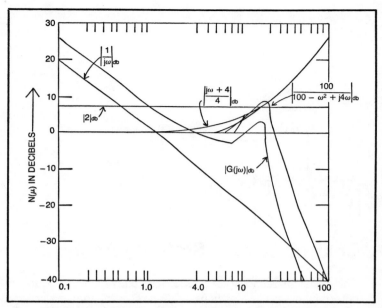

Fig. 4-24. Constructing the amplitude ratio curve for $G(s) = 50(s+4)/s(s^2+4s+100)$.

4-21. The quadratic term that contains a second-order term is sketched by using the normalized curves of Fig. 4-22. This second-order term has $\omega_n^2 = 100$ or $\omega_n = 10$ a damping of $\zeta = 0.2$, whose calculation is shown below.

$$2\zeta\omega_n = 4$$

$$\zeta = 4/2\omega_n = 4/(2)(10) = 4/20$$

$$\zeta = 0.2$$

At $\mu = 1$ in Fig. 4-22, this corresponds to $\omega = \mu\omega_n = (1)(10) = 10$ in Fig. 4-24. Hence, the second-order is sketched on Fig. 4-24 from Fig. 4-22 when $\zeta = 0.2$. Next, the composite amplitude curve is drawn as the sum of the four individual curves and is shown as the darker graph in Fig. 4-24.

It is possible to approximate frequency response curves for a transfer function of any order simply be decomposing it into the product of simple first-order and second-order functions. The asymptotic approximations made on the Bode plot require much less time than any other method that might be used to compute the frequency response of filter transfer functions.

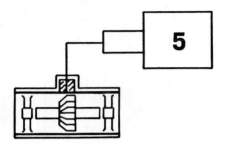

Control System Stability Analysis

The transient response of a feedback control system is uniquely determined by the locations of the roots of the characteristic equation in the s-plane. The characteristic equation of a control system is defined when the denominator of the transfer function is set equal to zero. Figure 5-1 illustrates some of the typical transient responses corresponding to the location of poles (roots of the characteristic equation) when located in the s-plane.

It is apparent from Fig. 5-1 that if any one of the real roots is positive—that is, located in the right half of the s-plane—its corresponding exponential term in the transient response will increase monotonically with time; in this condition, the system is said to be unstable. Similarly, a pair of complex conjugate roots with positive real parts will correspond to a sinusoidal oscillation with increasing amplitude. Therefore, for a stable response, the roots of the characteristic equation should not be found in the right half of the s-plane. Roots that are on the imaginary axis correspond to systems with sustained constant amplitude oscillations. The effect on the shape of the exponential and damped sinusoidal responses by various root locations in the s-plane is clearly shown in the four step responses of Fig. 5-1.

Basically, the stability and design of feedback control systems can be regarded as a problem of arranging the location of the characteristic equation roots in such a way that the corresponding system will perform according to the prescribed specifications, the most important requirement is that the system must

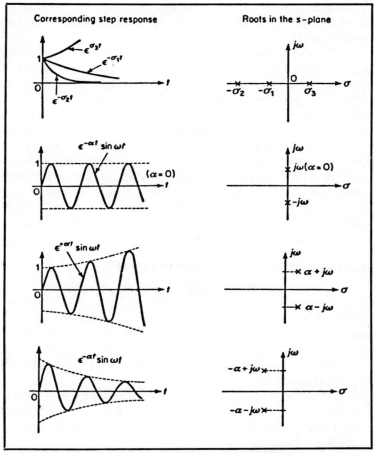

Fig. 5-1. Response comparison for various root locations in the s-plane.

be stable at all times. In other words, a system is defined as stable if the output response to any input disturbance is finite. This implies that all the roots of the characteristic equation must be located in the left half of the s-plane. Roots in the right half of the plane give rise to transients that tend to diverge from the steady state. The system is then said to be *unstable*.

If any physical output quantity of a control system, such as position, velocity, force, voltage, temperature, etc., reaches a very large magnitude as time goes by, the control system is unstable. This means that no physical system can respond in an ever-increasing manner, and that the response of an actual system must soon reach a level and remain at that level or be conditionally stable by oscillating back and forth between two limiting levels.

The linear stability analysis is not sufficiently comprehensive to allow us to compute the exact response of actual (nonlinear) systems which are unstable; however, this detracts little from the usefulness of linear stability analysis. The importance of linear stability analysis is in predicting whether unstable motions will occur and in showing how to avoid such motions, rather than in computing the exact form of such motions when they do occur. Thus, the stability analysis developed in this chapter is applicable to real systems which are linear for small changes in the physical variables. Many important control systems are of this class. Some important systems are not, though, and the materials in this chapter is consequently limited in its application to some control systems.

ROUTH-HURWITZ STABILITY DETERMINATION

One approach to the stability determination problem is to derive a transfer function for each of the system components which relates the physical properties to the poles and zeros. If the system transfer function is known, then the stability can be determined by inspection, because the system will be unstable if any poles of the system transfer function are in the right-half s-plane of the $j\omega$ axis.

The basic block diagram of a feedback control system was developed in Chapter 2 and is shown in Fig. 5-2. The closed loop transfer T(s) for Fig. 5-2 was given in Chapter 2 and is repeated in Equation 5-1.

$$T(s) = \frac{G(s)}{1 + G(s)H(s)}$$ **Equation 5-1**

If the individual forward path transfer function, G(s), and the feedback path transfer function, H(s) are defined as follows, Equation 5-1 is rewritten into a more convenient form to determine stability.

$$G(s) = \frac{N_1(s)}{D_1(s)}$$ **Equation 5-2**

$$H(s) = \frac{N_2(s)}{D_2(s)}$$ **Equation 5-3**

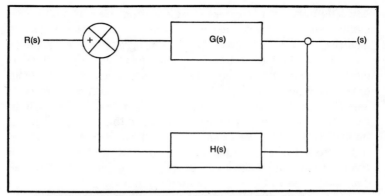

Fig. 5-2. Basic block diagram of a feedback control system.

Substituting equations 5-2 and 5-3 into equation 5-1 yields the following equation:

$$T(s) = \frac{N_1(s)\, D_2(s)}{N_1(s)\, N_2(s) + D_1(s)\, D_2(s)}$$

Equation 5-4

In this equation, the zeros of T(s) are determined when $N_1(s) = 0$ and $D_2(s) = 0$. Hence, there is a method of determining the zeros of the system transfer function if the transfer functions of the forward path and feedback of the control system are known. In order words, the overall system transfer function of the control system doesn't need to be known to find the zeros of the system transfer function. The poles of the system transfer function can be found in a similar manner.

The characteristic equation of the system transfer function determines the poles for the system. From equation 5-4, if you let the denominator of T(s) equal zero, this equation is called the *characteristic equation* for the system. The poles of T(s) occur at the values of *s* for which the following equation is satisfied.

$$N_1(s)\, N_2(s) + D_1(s)\, D_2(s) = 0$$

Equation 5-5

The determination of stability of a control system is reduced to knowing the transfer functions of the forward path and feedback of the control system, formulating the polynomial defined by equation 5-5, which has terms in *s*, and factoring the polynomial. The polynomial has real coefficients that are algebraic combinations of physical parameters, all having real magnitudes.

The polynomial expressed by equation 5-5 with real coefficients can be factored in several ways. The *least* satisfactory factoring is by long division. The programming of a digital computer to find the roots is the *most* satisfactory method of factoring polynomials. And in between are a number of methods of factoring polynomials by brilliant persons, such as A. Hurwitz and Edward Routh. The Routh-Hurwitz rules, which will be referred to as the *R-H rule*, is described in this section. Another method called the *root locus*, is a graphical procedure that not ony factors the polynomial but shows how the root locations change when one of the coefficients in the polynomial is varied. The root locus will be discussed later in this chapter.

It has been established that the problem of determining the stability of a linear system is one of finding the roots of the characteristic equation. However, for polynomials of the third order or higher, the task of finding the roots is tedious and time consuming. It is therefore desirable to use an alternate method so that the system stability can be determined without actually solving for the roots of the characteristic equation.

The factoring procedure of R-H has led to an R-H test to a polynomial which determines how many roots of the characteristic equation lie in the right-half s-plane. Remember, though, the R-H test is not actually a factoring method. We will be applying the R-H test to the characteristic equation, since it is easier than factoring the polynomial characteristic equation. First, equation 5-5 will be rewritten so that the R-H test of this equation can be shown.

$$D_T(s) = N_1(s) \, N_2(s) + D_1(s) \, D_2(s) = 0 \qquad \textbf{Equation 5-6}$$

Equation 5-6 can be expressed as the following polynomial:

$$D_T(s) = a_0 s^n = a_1 s^{n-1} + a_2 s^{n-2} + ... + a_{n-1} s^1 + a_n s^0 \quad \textbf{Equation 5-7}$$

The first step in the simplification of the R-H test is to arrange the polynomial coefficients of equation 5-7 into two rows as shown below.

$$
\begin{array}{c|l}
s^n & a_0 \; a_2 \; a_4 \; ..... \; a_n \\
s^{n-1} & a_1 \; a_3 \; a_5 \; ..... a_{n-1}
\end{array}
$$

These two rows are the start of an array which will be described by an example for a sixth-order (n = 6) polynomial system. A complete array in the R-H test for a polynomial of the sixth order will next be formed.

s^6	a_0	a_2	a_4	a_6
s^5	a_1	a_3	a_5	0
s^4	$\dfrac{a_1 a_2 - a_3 a_0}{a_1} = A$	$\dfrac{a_1 a_4 - a_0 a_5}{a_1} = B$	$\dfrac{a_1 a_6 - a_0(0)}{a_1} = a_6$	0
s^3	$\dfrac{A a_3 - a_1 B}{A} = C$	$\dfrac{Aa_5 - a_1 a_6}{A} = D$	$\dfrac{A(0) - a_1(0)}{A} = 0$	0
s^2	$\dfrac{CB - AD}{C} = E$	$\dfrac{Ca_6 - A(0)}{C} = a_6$	$\dfrac{C(0) - A(0)}{C} = 0$	0
s^1	$\dfrac{ED - Ca_6}{E} = F$	0	0	0
s^0	$\dfrac{Fa_6 - E(0)}{F} = a_6$	0	0	0

The array terminates when all the remaining terms in the first column of the array are computed to be zero. Next, inspect the numbers in the first column to determine if the system is stable or unstable. If the algebraic signs of all terms in the first column of the array are alike, all the roots of the characteristic equation lie in the left half of the s-plane, and the system if stable. If the signs of all the numbers in the first column are not alike, the system is unstable, and there are as many roots in the right-half s-plane as there are changes in algebraic sign. The following examples will illustrate the R-H test.

□ **Example 5-1:** Test the following polynomial for stability using the R-H test: $T(s) = s^3 - 4s^2 - 5s + 6$.

The polynomial given has negative coefficients; thus, by inspection you know there are positive real roots. The R-H test should therefore illustrate an unstable system. The array for the R-H test is the following:

s^3	1	-5
s^2	-4	6

$$s^1 \quad \frac{(-4)(-5)-(1)(6)}{(-4)} = -3.5 \qquad 0$$

$$s^0 \quad \frac{(-3.5)(6)-(-4)(0)}{-3.5} = 6 \qquad 0$$

In this array, there are two sign changes in the first column. The polynomial has then two roots located in the right-half of the s-plane. Thus, the system is unstable. If the polynomial is factored, the two unstable roots are found to be s = 2 and s = 3.

The next example considers a fourth-order polynomial.

□ **Example 5-2:** Test the following polynomial for stability by using the R-H test: $T(s) = 2s^4 + s^3 + 3s^2 + 5s + 10$.

The array for the R-H test is the following:

s^4	2	3	10
s^3	1	5	0
s^2	$\dfrac{(1)(3) - (2)(5)}{1} = -7$	$\dfrac{(1)(10) - (2)(0)}{1} = 10$	0
s^1	$\dfrac{(-7)(5) - (1)(10)}{-7} = 6.43$	0	0
s^0	10	0	0

In this array, two sign changes are in the first column. The polynomial has then two roots located in the right-half of the s-plane. Thus, the system is unstable.

In applying the R-H test, two special cases occur when computing the array. The first special case occurs when in any one row of the array the first number of that row is zero, while the other numbers are not. The second special case occurs when in any row all the numbers are zero. These two cases will be illustrated separately.

When the first element in any row of the R-H test array is zero, while the other elements are not, leads to the fact that the elements in the next row become infinite, the R-H test breaks down. To restore the missing power of s and eliminate a row beginning with the number zero, simply multiply the polynomial by the factor, (s + a), where *a* is any positive real number, and carry on the R-H test as usual. The following example will illustrate this special case.

☐ **Example 5-3:** Test the following polynomial for stability by using the R-H test. $T(s) = s^3 - 3s + 2$

Because the coefficient of the s^2 term is zero, we know by inspection that there must be at least one root of the polynomial which is located in the right-half s-plane. The following is the tabulation of R-H array for the given polynomial:

s^3	1	0
s^2	0	2
s^1	∞	
s^0	∞	

In this array, because of the zero in the first element of the second row, the first element of the third row is infinite. To correct this situation, simply multiply the given polynomial, $T(s)$, by factor $(s + 3)$. The result of this multiplication for the new polynomial, $T'(s)$ is the following:

$$T'(s) = (s + 3)T(s) = s^4 + 3s^3 - 3s^2 - 7s + 6$$

The array for the R-H test of the new polynomial, $T'(s)$, follows.

s^4	1	-3	6
s^3	3	-7	0
s^2	$\dfrac{(3)(-3) - (1)(-7)}{3} = \dfrac{-2}{3}$ 6	0	
s^1	$\dfrac{(-\frac{2}{3})(-7) - (3)(6)}{(-\frac{2}{3})} = +20$ 0	0	
s^0	6	0	0

In this array, there are two changes in sign in the first column. There are therefore two roots of the polynomial in the right-half s-plane, indicating that the system is unstable.

For the second special case, when all the elements in one row of the R-H test array are zero, there are pairs of real roots with opposite signs, pairs of conjugate roots on the imaginary axis, or both; or conjugate roots forming a quadrate in the s-plane. The equation corresponding to the coefficients just above the row of zeroes is called the auxiliary equation. The order of the auxiliary equation is always even, and it indicates the number of root pairs that are equal in magnitude but opposite in sign. For example, if the auxiliary equation is of the second order, there are two equal and opposite roots. For a fourth-order auxiliary equation, there must be two *pairs* of equal and opposite roots. All these roots with equal magnitude can be obtained by solving the auxiliary equation. Again, the R-H test breaks down—in this case, because of the row of zeros. To correct this situation, simply take the first derivative of the auxiliary equation with respect to s, replace the row of zeros with the coefficients of the resultant equation obtained by taking the derivative of the auxiliary equation, and carry on with the R-H test. The following examples will illustrate the technique of the second special case.

□ **Example 5-4:** Test the following polynomial for stability using the R-H test: $T(s) = s^4 + s^3 - 3s^2 - s + 2$.

The array for the R-H test is the following:

s^4	1	-3	2
s^3	1	-1	0
s^2	$\dfrac{-3+1}{1} = -2$	2	Auxiliary equation coefficients
s^1	0	0	

Because the s^1 row contains all zeros, the R-H test breaks down. The auxiliary equation is obtained by using the numbers contained in the s^2 row as the coefficients of the auxiliary equation. Therefore, the auxiliary equation can be written as follows.

$$A(s) = -2s^2 + 2$$

$$\frac{dA(s)}{ds} = -4s$$

The row of zeros in the initial array for the given polynomial are replaced by the coefficients of the derivative of the auxiliary equation. The new array follows. From this new array the R-H test continues.

s^4	1	-3	2
s^3	1	-1	
s^2	-2	2	
s^1	-4	0	The coefficients of dA(s)/ds
s^0	2	0	

Because there are two changes in sign in the numbers of the first row of the new array of the R-H test, two roots of the original polynomial are located in the right half of the s-plane; the system is therefore unstable.

Another example is furnished for this special case so that you can become familiar with the details of the R-H test.

□ **Example 5-5:** Test the following polynomial for stability using the R-H test: $T(s) = s^6 + s^5 + -2s^4 - 3s^3 - 7s^2 - 4s - 4$.

The array for the R-H test is shown below.

s^6	1	-2	-7	-4
s^5	1	-3	-4	
s^4	$\dfrac{1(-2) - 1(-3)}{1} = 1$	$\dfrac{-7 + 4}{1} = -3$	-4	
s^3	0	0	0	

Because the s^3 row contains all zeros, the R-H test breaks down. The auxiliary equation is obtained by using the numbers contained in the s^4 row as the coefficients of the auxiliary equation. The auxiliary equation can then be written as follows.

$$A(s) = s^4 - 3s^2 - 4$$

$$\frac{dA(s)}{ds} = 4s^3 - 6s$$

The row of zeros in the initial array for the given polynomial are replaced by the coefficients of the derivative of the auxiliary equation. The new array follows. From this new array the R-H test continues.

s^6	1	-2	-7	-4
s^5	1	-3	-4	
s^4	1	-3	-4	
s^3	4	-6	0	The coefficients of $\dfrac{dA(s)}{ds}$
s^2	$\dfrac{4(-3) - 1(-6)}{4} = -1.5$	-4	0	

111

$$s^1 \quad \frac{-9 + 16}{-1.5} = -16.7 \qquad 0$$

$$s^0 \qquad -4$$

In Example 5-5 because there is only one change in sign in the first column of the R-H tabulation, the polynomial has one root in the right-half s-plane.

A frequent use of the R-H test is to determine the condition of stability of a linear feedback control system. The next example will illustrate how to determine the range of a system parameter to determine stability using the R-H test.

□ **Example 5-6:** A servo system has the following characteristic equation: $T(s) = s^3 + 34.5s^2 + 7500s + 7500K_1$. Using the R-H test, determine the range of parameter K_1 for which the closed loop control system is stable.

The array for the R-H test is as follows.

s^3	1	7500
s^2	34.5	$7500K_1$
s^1	$\dfrac{(258,750 - 7500K_1)}{34.5}$	0
s^0	$7500K_1$	

Remember that for the control system to be stable, the terms in the first column of the R-H array must be positive. This means the following conditions must be met.

$$\frac{(258,750 - 7500K_1)}{34.5} > 0$$

and also:

$$7500K_1 > 0$$

From the conditions in these equations the conditions of stability for parameter K_1 which gives us a range of K_1 can be calculated, as shown below.

$$0 < K_1 < 34.5$$

THE ROOT LOCUS METHOD OF DETERMINING STABILITY

A method widely used to determine stability in control systems was developed by W. R. Evans. It is called the *root locus method.* The reason for its popularity is that it shows graphically how the locations of the poles of a system transfer function change as one of the physical parameters of the system is changed. It is then possible for a designer to determine the proper values for the physical parameters on the basis of the desired dynamic response of the closed loop system. Furthermore, it shows in a very direct manner whether the desired dynamic performance can be achieved simply by adjusting the parameter values of a given system. If the required performance cannot be achieved, the root locus method often indicates the manner in which the system should be redesigned to meet the performance specifications.

In the block diagram of Fig. 5-2, the closed loop transfer function can be written as follows:

$$T(s) = \frac{G(s)}{1 + G(s)\,H(s)} \qquad \textbf{Equation 5-8}$$

Next, it's desirable to determine the poles of the closed loop transfer function since the poles characterize the response of the control system. Therefore, the equation to be solved is the following and is called the characteristic equation;

$$1 + G(s)H(s) = 0 \qquad \textbf{Equation 5-9}$$

Equation 5-9 can be solved numerically to determine the poles. However, what is wanted is the dependence of the poles on a system parameter, such as the system gain. This is necessary since the parameter might not be known exactly, or because you might wish to change the parameter and want to observe how the parameter change will affect location of the poles. The root locus method basically gives the closed loop poles in relation to the open loop poles and zeros, and a parameter K, which is the parameter of interest (usually gain). The root locus method will be described by the following set of rules, each of which is illustrated by an example.

☐ Rule 1:

In order to construct the root loci, obtain the characteristic equation, and rearrange it in the following form.

$$1 + \frac{K(s - z_1)\,(s - z_2)\,\ldots\ldots\,(s - z_m)}{(s - p_1)\,(s - p_2)\,\ldots\ldots\,(s - p_n)} = 0 \qquad \textbf{Equation 5-10}$$

Where K is the parameter of interest in the control system usually gain, z_m is the mth zero of the characteristic equation, and p_n is the nth pole of the characteristic equation.

The root locus plot starts at each pole of the characteristic equation, and the root locus plot terminates on the zeros of the characteristic equation.

Example 5-7: The forward path transfer function, G(s), is

$$\frac{K}{s(s + 1),}$$

and the feedback path transfer function, H(s), is

$$\frac{(s + 2)}{(s + 3)\,(s + 4)}.$$

Find the poles and zeros of the characteristic equation, and plot the poles and zeros in the s-plane.

First, construct the characteristic equation.

$$1 + G(s)H(s) = 1 + \frac{K(s + 2)}{s(s + 1)\,(s + 3)\,(s + 4)} = 0$$

Equation 5-11

Then list the poles and zeros from this equation.

- ☐ $s = -2$ is a zero.
- ☐ $s = 0$ is a pole.
- ☐ $s = -1$ is a pole.
- ☐ $s = -3$ is a pole.
- ☐ $s = -4$ is a pole.

The poles and zeros are plotted in s-plane of Fig. 5-3.

☐ Rule 2

This rule tells us the starting and termination points of the root loci of the characteristic equation. For all real control systems, the number of open loop/poles is greater or equal to the number of zeros. This means the characteristic equation defined by Equation 5-10 must have n equal to or greater than m ($n \geqslant m$). The root loci end at the zeros of the characteristic equation, and the root loci start at the poles of the characteristic equation.

Next, solve for the value of K in the characteristic equation at the pole and zero values. From Equation 5-10, K is defined as follows.

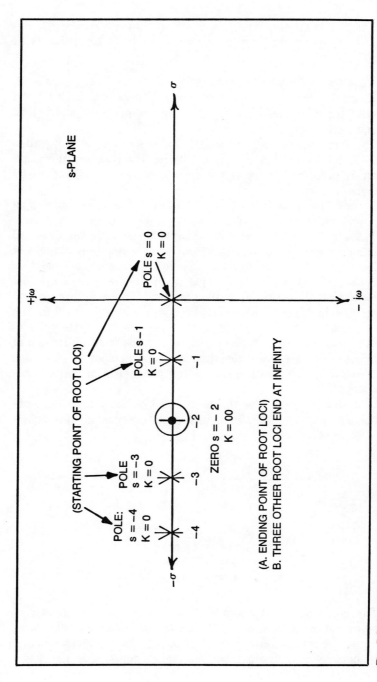

Fig. 5-3. An s-plane plot of poles and zeros for Example 5-7.

115

$$K = \frac{-(s - p_1)(s - p_2) \dots (s - p_n)}{(s - z_1)(s - z_2) \dots (s - z_m)}$$ **Equation 5-12**

If the pole values are substituted one at a time in Equation 5-12, K equals 0 at each of the pole values. Therefore, the root loci start for K equals 0 at the pole values of the characteristic equation.

If the zero values are substituted one at a time in Equation 5-12, the value of K would be infinity. The root loci terminate for K equal to infinity at the zero values of the characteristic equation.

For the root locus plot, there are n branches (number of poles) and m (number of zeros) which will terminate at a zero. Thus, for the characteristic equation to be rational, (n − m) branches must terminate at infinity along asymptotes of the root locus plot. The characteristic equation of Example 5-7 has four poles is 4 and one zero. This means that one branch of the root locus plot will terminate at the zero, and three will terminate at infinity. In other words, one finite zero exists at $s = -2$ for the equation of Example 5-7. Because there are four finite poles of the same equation, three $(4 - 1 = 3)$ zeros must be located at infinity. For a rational function, the total number of poles and zeros must be equal if the poles and zeros at infinity are counted.

□ **Example 5-8:** For the following functions, find the finite poles and zeros, the infinite zeros, and the number of branches in the root locus plot:

a. $1 + \dfrac{K(s + 1)}{s(s + 2)(s + 3)}$ b. $1 + \dfrac{K}{s^2(s + 2)(s + 3)}$

a. $s = -1$ is a zero
$s = 0$ is a pole
$s = -2$ is a pole
$s = -3$ is a pole

The poles and zero listed are finite. Because the number of poles and zeros must be equal for a rational function, there must be two additional zeros located at infinity for the given function to have exactly three poles and three zeros. For the given equation, there are three finite poles (n = 3) and one finite zero (m = 1). The number of branches in the root locus is equal to n; that is, the number of branches is equal to three in this example. However, only one branch will terminate at the zero, since m = 1. This means that (n − m) branches must terminate at infinity. In this example, (n −m) = (3 − 1), or two branches must terminate at infinity.

b. s = 0 is double pole
s = −2 is a pole
s = −3 is a pole

The poles listed are finite. Because the number of poles and zeros must be equal for a rational function, there must be four additional zeros located at infinity in order for the given function to have exactly four poles and four zeros. The number of branches in the root locus is equal to n, four branches exist in the root locus of this example. This means that (n − m) = (4 − 0), or four branches must terminate at infinity.

Before leaving Rule 2, let's summarize the details illustrated while describing the rule. The summary of Rule 2 is:

☐ From the characteristic equation, find the finite poles, n, and the finite zeros, m.

☐ Calculate the zeros located at infinity, which is equal to (n − m).

☐ The branches of the root locus will terminate at a zero.

☐ The branches of the root locus will start at a pole.

☐ At a zero, the value of K is infinity.

☐ At a pole, the value of K is zero.

☐ Rule 3

The root locus plot is symmetrical with resepct to the real axis. For rational functions, the complex roots must appear in complex conjugate pairs, hence the branches of the root locus plot will be symmetrical with respect to the real axis of the s-plane.

☐ Rule 4

This rule tells us how to determine the asymptotes of the root locus plot. The asymptotes in the root locus plot exist for the zeros located at infinity. In other words, the asymptotes of the root locus plot must be determined for the (n − m) zeros located at infinity, because the branches of the root locus plot will never reach infinity, but they will approach the asymptotic value of the root locus plot at the infinity zero location. For large values of s, therefore, the root loci are asymptote to straight lines with angles measured from the real axis, and the angles given by the following equation:

$$\beta = \frac{180° \ (2 \ k + 1)}{n - m} \qquad \textbf{Equation 5-13}$$

where β = the angle between the asymptote and the real axis, n = the number of finite poles in the characteristic equation, m = the number of finite zeros in the characteristic equation, and k = A series of numbers, such as 0, 1, 2, ... (n − m − 1). As can be seen k begins at 0 and ends at (n − m − 1).

All the asymptotes for a given characteristic equation intersect the real axis at a point, σ (sigma). The point, σ, is given by the following equation.

$$\sigma = \frac{(p_1 + p_2 + ... + p_n) - (z_1 + z_2 + ... + z_n)}{n - m} \qquad \textbf{Equation 5-14}$$

□ **Example 5-9:** For the characteristic equation given in Example 5-7, find the angles of the asymptotes for the zeros located at infinity, find the intersection point of the asymptotes on the real axis, and draw an s-plane plot of the asymptotes associated with the root locus plot of the given characteristic equation.

First we will list the information needed from example 5-7 in order to solve this problem.

$p_1 = 0$, $p_2 = -1$, $p_3 = -3$, $p_3 = -4$, $z_1 = -2$.
n = 4, m = 1, (n − m) = 4 − 1 = 3 zeros at infinity, which means that three asymptotes exist.
k = 0, 1, 2.

To find the angles of the asymptotes, employ equation 5-13 as follows:

$$B_0 \ = \ \frac{180° \ (2 \times 0 + 1)}{3} \ = \ 60°$$

$$B_1 \ = \ \frac{180° \ (2 \times 1 + 1)}{3} \ = \ 180° = 180°$$

$$B_2 \ = \ \frac{180° \ (2 \times 2 + 1)}{3} \ = \ 180° \ \frac{(5)}{3} \ = 300° = -60°$$

118

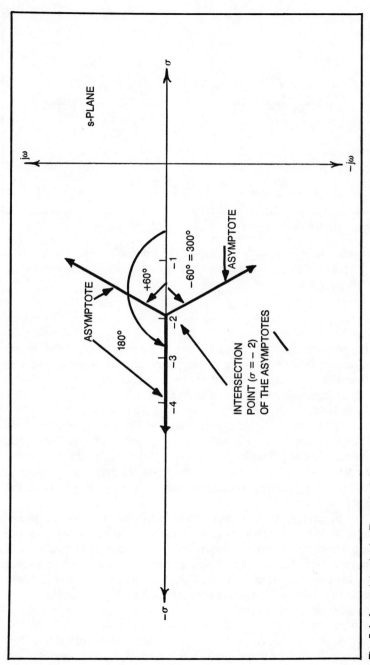

Fig. 5-4. Asymptote plot for Example 5-9.

119

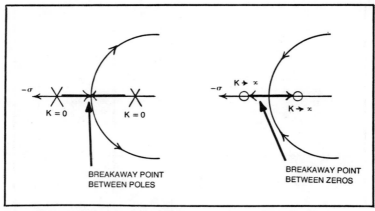

Fig. 5-5. Breakaway points on the real axis of the s-plane.

Thus, the angles we will consider will be 180°, +60°, and −60°. To find the intersection point of the asymptotes on the real axis, use equation 5-14 as follows:

$$\sigma = \frac{(0 - 1 - 3 - 4) - (-2)}{3} = -2$$

Be warned that the intersection point of the asymptotes is not always a finite zero point of the characteristic equation, as was found above. Next we can draw the asymptotes of the root locus plot in the s-plane, as shown in Fig. 5-4.

☐ Rule 5:

When the characteristic equation contains multiple roots (roots of an order higher than one; roots of two are quite common), the points in the s-plane where the multiple roots are found are called the *breakaway points* of the root locus diagram. Rule 5 will illustrate the techniques of breakaway points in the root locus method.

Figure 5-5A illustrates a case in which two separate loci of a root locus diagram meet at a point on the real axis in the s-plane and then break away from the real axis as the value of K is increased further. The point at which the two root loci meet and break away is a breakaway point; in this case, it represents a double root of the characteristic equation. Figure 5-5B shows a similar situation where two loci of complex roots break away at a point on the real axis in the s-plane and then approach the two zeros. In general, a breakaway point may involve more than two root loci. For example, Fig. 5-6 shows that four separate loci meet at a point on the real axis and then depart in different directions. Also, there might be more than one breakaway point for a root locus diagram, but because of the conjugate symmetry of the root loci, the breakaway points must either lie on the real axis or occur in complex conjugate pairs.

In Fig. 5-5 the root loci are shown to breakaway on the real axis at angles of 180° apart, while in Fig. 5-6 the four loci depart at angles of 90° apart. A general equation concerning the angles between the loci at a breakaway point can be written as follows:

$$\Theta_{BA} = \frac{180°}{n} \qquad \textbf{Equation 5-15}$$

where Θ_{BA} = the angle the root loci must approach and leave a breakaway point on the real axis, and n = number of root loci approaching and leaving the breakaway point.

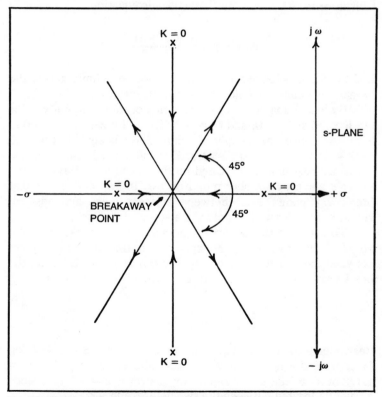

Fig. 5-6. Breakaway point of four separate loci on the real axis.

121

□ Rule 5A

Rule 5A concerns itself with breakaway points due to poles and zeros on the real axis. If only the real poles and zeros of the characteristic equation are considered, and −a is assumed to be the breakaway point on the real axis, the value of a is determined by the following equation.

$$\sum \frac{1}{Zi-a} - \sum \frac{1}{Pj-a} = \sum \frac{1}{a-Zi} - \sum \frac{1}{a-Pj} \qquad \textbf{Equation 5-16}$$

| Zeros to left of −a | Poles to left of −a | Zeros to right of −a | Poles to left of −a |

Although Equation 5-16 appears complicated, the following example will show that the application of the above equation is simple.

□ **Example 5-10:** For the characteristic equation,

$$T(s) = 1 + \frac{K(s + 4)}{s\,(s + 2),}$$

find the finite poles and zeros, the number of infinite zeros, the breakaway points, and plot the root locus diagram.

By inspection of the characteristic equation, the finite poles are located at $p_1 = 0$, and $p_2 = -2$. The finite zero is located at $z_1 = -4$. Hence, the number of infinite zeros is equal to (n − m) = (2 − 1) = 1. One zero is located at infinity. All the poles and zeros are real numbers located on the negative real axis of the s-plane, and they are shown on Fig. 5-7. In Fig. 5-7, notice that breakaway points exist between adjacent poles and adjacent zeros, but not between a pole and a zero.

To calculate the breakaway points, let the breakaway point between the two poles be at −a and the breakaway point between the two zeros be at −b. You can therefore calculate the breakaway point, −a, by applying equation 5-16 as follows:

$$\frac{1}{4 - a} - \frac{1}{2 - a} = \frac{-1}{a - 0}$$

Rearranging the terms in this equation, $a^2 - 8a + 8 = 0$. Solving for a by factoring or using the quadratic equation yields $a_1 = 1.172$ or $a_2 = 6.828$. The significant answer is a = −1.172, since a must be located between the two poles 0 and −2.

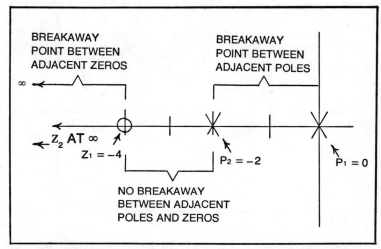

Fig. 5-7. Location of zeros and poles and breakaway points for Example 5-10.

Next, calculate the breakaway point, $-b$, by applying equation 5-16 as follows:

$$\frac{1}{\infty} = \frac{1}{b-4} - \frac{1}{b-2} - \frac{1}{b-0}$$

$$0 = \frac{1}{b-4} - \frac{1}{b-2} - \frac{1}{b}$$

Rearranging the terms in this equation, $b^2 - 8b + 8$. Solving for b by using the quadric equation yields $b_1 = 1.172$ or $b_2 = 6.828$. The significant answer is $b = -6.828$, since b must be located between the two zero values -4 and infinity. Notice that the answers for both a and b are the same. This is merely a coincidence that will not happen most of the time.

When all the poles and zeros of the characteristic equation of the transfer function are located on the negative real axis, all breakaway points will be located on the negative real axis. Figure 5-8 shows the pole zero plot and the root locus breakaway diagram.

As shown in Fig. 5-8, if the root locus lies between two adjacent poles on the real axis, there is at least one breakaway point ($a = -1.172$). Similarly, if the root locus lies between two adjacent zeros on the real axis, there is at least one break-in point (b = -6.828).

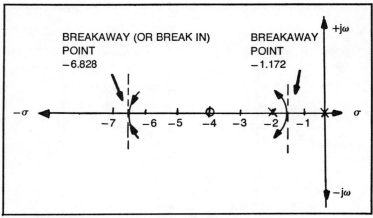

Fig. 5-8. Partial solution to Example 5-10.

The astute reader will notice that the breakaway and the break-in points are equidistant from the finite pole, -4. In other words, the breakaway point, $a = -1.172$, and the break-in point, $b = -6.828$, are exactly 2.828 units from the pole, -4. This means a root locus diagram can be drawn using the pole, -4, as the center of a circle and the circle having tangents at the breakaway and break-in points as shown in Fig. 5-9.

☐ Rule 5B

Rule 5B concerns itself with the breakaway points due to complex poles and zeros. The contribution to the breakaway and break-in points on the real axis from the complex poles and zeros of the characteristic equation is determined from the following equation:

$$\sum \frac{2(\alpha_i - a)}{(\alpha_i - a)^2 + \beta_i^2} \quad - \quad \sum \frac{2(\alpha_j - a)}{(\alpha_j - a)^2 + \beta_j^2} \quad = \quad \frac{2(a - \alpha_i)}{(a - \infty_i)^2 + \beta_i^2}$$

Complex   Complex   Complex
Zeros to   Poles to   Zeros to
left of $-a$ left of $-a$ right of $-a$

$$- \quad \sum \frac{2(a - \alpha j)}{(a - \alpha j^2) + \beta_j^2} \qquad \textbf{Equation 5-17}$$

Complex
Poles to
right of $-a$

Although Equation 5-17 appears complicated, the following example will show that the application of the above equation is simple.

□ **Example 5-11:** For the characteristic equation,

$$1 + \frac{K(s + 2)}{s^2 + 2s + 2}$$

find the finite poles and zeros, the number of infinite zeros, the breakaway (break-in) point on the real axis, and plot the root locus diagram.

In order to locate the finite poles of the given equation, employ the quadratic equation as follows.

$$p_1 \text{ and } p_2 = \frac{-2 \pm \sqrt{2^2 - (4)(1)(2)}}{2(1)} = \frac{-2 \pm \sqrt{-4}}{2} = -1 \pm j1$$

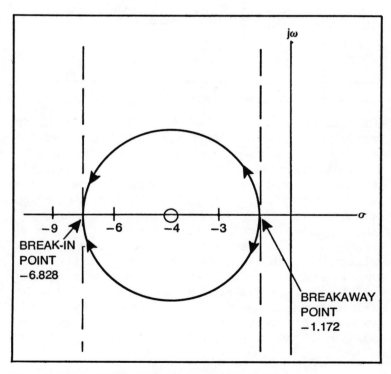

Fig. 5-9. Root locus diagram for Example 5-10.

125

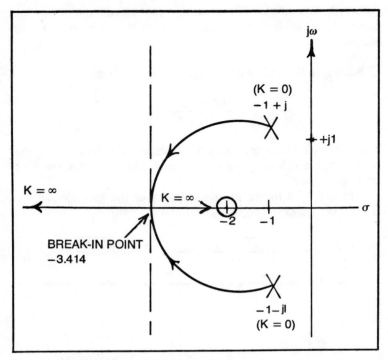

Fig. 5-10. Root locus diagram for Example 5-11.

Then you can say that $p_1 = -1 + j1$ and $p_2 = -1 -j1$. By inspection the finite zero is located at $z_1 = -2$. The number of infinite zeros is therefore equal to $(n - m) = (2 - 1) = 1$. Because there are two adjacent zeros on the real axis, a break-in point must be found between the two zeros. The break-in point can be calculated by applying equation 5-17. First, you need the sum of complex zeros and poles to the left of the break-in point, but none exist; therefore, place a zero for the left side of equation 5-17. There is one complex pole to the right of the break-in point with $\alpha_j = 1$ and $\beta_j = 1$. There are no complex zeros to the right of the break-in point so that term in equation 5-17 is zero. There is one real number zero to the right of the break-in; therefore, you must employ Equation 5-16 along with Equation 5-17 to calculate the break-in point. The combination of Equation 5-16 and Equation 5-17 is as follows:

$$\frac{1}{a - 2} - \frac{2(a - 1)}{(a - 1)^2 + 1^2} = 0$$

126

Rearranging the terms in the equation, $a^2 - 4a + 2 = 0$. The break-in point, a, in the above is found to be 3.414. Next, plot the root locus diagram as shown in Fig. 5-10.

□ Rule 5C

Rule 5C concerns itself with the breakaway point not on the real axis. Equations 5-16 and 5-17 can be extended to compute the breakaway (break-in) points which are not located on the real axis of the s-plane. Consider the root loci given in Fig. 5-11; the breakaway point on the real axis is readily determined to be -2. However, there are two conjugate breakaway points on the complex branches of the root loci. Because of the symmetry of the root loci, it is necessary to determine only one of the two points. If the axis on which the two breakaway points lie is interpreted as the real axis, Equations 5-16 and 5-17, as previously given, can be applied.

In Fig. 5-12, the pole at $-2 + j4$ is arbitrarily chosen as the new orgin, and the new set of coordinates is as indicated. If the

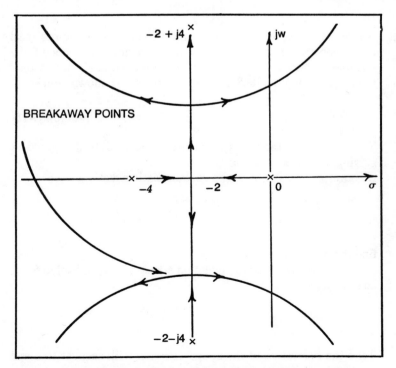

Fig. 5-11. Root loci with breakaway points not located on the real axis.

127

location of the breakaway point in the upper half of the s-plane is represented as $-a$ measured from the new origin O', a is determined by the equation:

$$\frac{-1}{(8-a)} - \frac{2(4-a)}{(4-a)^2 + 2^2} = \frac{-1}{a}$$

Solving for a in the equation yields a = 1.55. The breakaway points of the complex root loci are located at s = $-2 \pm j2.45$ and are shown in Fig. 5-12.

☐ Rule 5D

Rule 5D concerns itself with breakaway point computed by the analytical method. The breakaway points of a root locus diagram can also be computed by an analytical method. In this method, the characteristic equation is solved for parameter K as follows:

$$K = f(s)$$

where f(s) does not contain K, and the breakaway points (real and complex) of the root locus diagram are the roots of the equation obtained by taking the first derivative of K with respect to s, and setting the derivative equal to zero as shown below:

$$\frac{dK}{ds} = 0 \qquad \textbf{Equation 5-18}$$

The following example will illustrate the analytical method of solving for breakaway points.

☐**Example 5-12:** Given the characteristic equation,

$$1 + \frac{K(s+4)}{s(s+2)}$$

find the breakaway of this equation.

Set the characteristic equation equal to zero and solve for K as follows.

$$K = \frac{-s(s+2)}{(s+4)}$$

Take the derivative of the above equation with respect to s and set the derivative equal to zero.

$$\frac{dK}{ds} = \frac{-(2s+2)(s+4) - s(s+2)}{(s+4)^2} = 0$$

Simplifying the preceding equation obtains the following result:

$$s^2 + 8s + 8 = 0$$

The roots of the above equation represent the breakaway points, which are $s = -1.172$ and $s = -6.828$.

□ Rule 6

Rule 6 concerns itself with the calculation of K on the root loci. Once the root loci of the characteristic equation are constructed, the value of K at any point s_1 on the loci can be determined for the following equation:

$$K = \frac{1}{G(s_1) H(s_1)} \qquad \textbf{Equation 5-19}$$

Equation 5-19 can be evaluated either graphically or analytically. Usually, if the root locus plot is already constructed, the graphical method is more conveneint. For example, in Fig. 5-13 the value of K at the point s_1 on the root locus is given by K = (A)(B)(C)(D), where A, B, C, and D are the lengths of the vectors drawn from the open loop poles to the point s_1.

The value of K at the point at which the locus the locus intersects the imaginary axis is usually obtained by applying the Routh criterion to the characteristic equation.

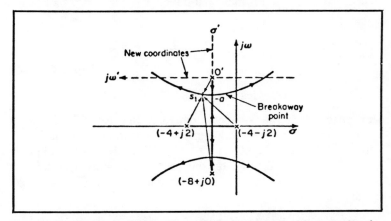

Fig. 5-12. A new set of coordinates assigned to the root locus diagram given in Fig. 5-11, in order that the complex breakaway points can be evaluated.

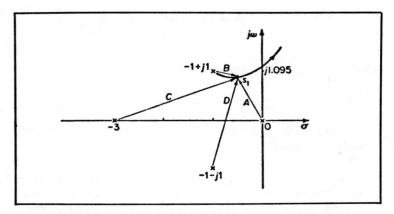

Fig. 5-13. Graphical method of evaluation the values of K on the root loci.

□ Rule 7

Rule 7 concerns itself with the intersection of the root loci with the imaginary axis. The root loci shown in Fig. 5-14 intersects the imaginary axis. The values of K and ω at the crossing point are determined by means of the Routh-Hurwitz criterion presented earlier in this chapter. An example will be shown later in this chapter.

□ Rule 8

Rule 8 concerns itself with the angles of departure from poles and the angles of arrival at zeros of the root loci. The angles of departure and arrival of the root loci can be determined readily from the following equation.

$$\underline{|G(s)\,H(s)} = \sum_{i=1}^{m} \underline{|s+zi} - \sum_{j=1}^{m+n} \underline{|s+pj} = (2k+1)\pi$$

Equation 5-20

↑ Sum of zero angles   ↘ Sum of pole angles

For instance, in the pole zero configuration given in Fig. 5-14, it is desired to determine the angle at which the root locus leaves the pole at $-1 + j1$. At point s_1, which is very close to the pole at $-1 + j1$, is selected on the root locus; since the point is assumed to be on the root locus, it must satisfy Equation 5-20. Thus, we have;

$$-(\Theta_{p1} + \Theta_{p2} + \Theta_{p3} + \Theta_{p4}) = (2k+1)180°$$

130

The Θs in this equation are measured as shown in Fig. 5-14. Therefore,

$$-135° - 90° - 26.6° - Θ_{p4} = (2k + 1)180°$$

$$Θ_{p4} = -251.6° - (2k + 1)180°$$

$$Θ_{p4} = -71.6°$$

The following example will illustrate the use of the rules presented in this section that are needed to construct a root locus diagram.

□ **Example 5-13:** The loop transfer functionof a feedback control system is given as follows:

$$G(s)H(s) = \frac{K(s + 3)}{s(s + 5)(s + 6)(s^2 + 2s + 2)}$$

Draw the root loci of the system from the rules of construction given in this section.

There are five poles for the given transfer function, which are $p_1 = 0$, $p_2 = -5$, $p_3 = -6$, $p_4 = -1 + j1$, and $p_5 = -1 - j1$. There is one finite zero located at $z_1 = -3$, and there are four poles located at infinity.

It is apparent, because the root loci must start at the poles and end at the zeros of G(s)H(s), there must be as many root locus as the larger value of zero or poles. Therefore, because

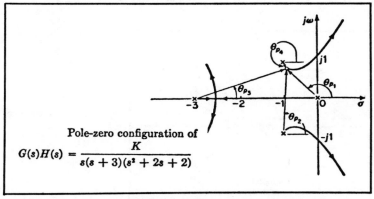

Fig. 5-14. Pole-zero configuration of G(s)H(s) = K/s(s + 3)(s² + 2s + 2).

there are five finite poles, there must be five root loci. The root loci must be symmetrical with respect to the real axis.

The angles of the asymptotes of the loci at infinity were given by Equation 5-13, and are the following:

$$\beta = \frac{180°(2k + 1)}{n - m} \quad \text{Equation 5-13}$$

where k in the example is k = 0, 2, 3, and n — m = 5 — 1 = 4. Remember that the last number for k is (n — m — 1) = 4 —1 = 3. Therefore, the four loci that terminate at infinity should approach infinity at the angles calculated below.

$$\beta_1 = \frac{180° (2 \times 0 + 1)}{4} = +45°$$

$$\beta_2 = \frac{180°(2 \times 1 + 1)}{4} = +135°$$

$$\beta_3 = \frac{180°(2 \times 2 + 1)}{4} = +235° = -135°$$

$$\beta_4 = \frac{180°(2 \times 3 + 1)}{4} = +315° = -45°$$

The asumptotes of root loci intersect at the point described by Equation 5-14. For this example, the point is given below:

$$\sigma = \frac{(0 - 5 - 6 - 1 + j1 - 1 - j1) - (-3)}{4} = -2.5$$

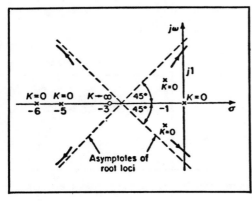

Fig. 5-15. Pole-zero plot.

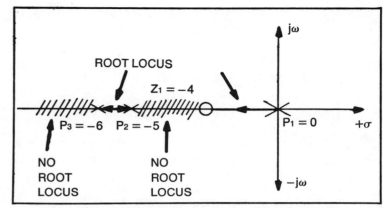

Fig. 5-16. Real axis critical points for root locus diagram of Example 5-13.

The results obtained in the last six steps are illustrated Fig. 5-15.

Loci are on the real axis between $s = 0$, and $s = -3$, $s = -5$, and $s = -6$. There are no root loci between $s = -3$ and $s = -5$, or between $s = -6$ and infinity. Critical points on the real axis for the root locus diagram are illustrated in Fig. 5-16.

The angles of departure of the root locus leaving the pole at $-1 + j1$ is determined by solving for Θ by using Equation 5-20 as follows:

$$\underset{\underline{s+3}}{26.6°} \; - \; (\underset{\underline{s}}{135°} \; + \; \underset{\underline{s+1+z1}}{90°} \; + \; \underset{\underline{s+5}}{14°} \; + \; \underset{\underline{s+6}}{11.4°} \; + \; \underset{\underline{s+1-j1}}{\Theta})$$

$$= (2k + 1)180°$$

From this equation, $\Theta = -43.8°$.

The interceptions of the root loci with the imaginary axis are determined by the Routh criterion. The characteristic equation of the system is found by letting $1 + G(s)H(s) = 0$:

$$s^5 + 13s^4 + 54s^3 + 82s^2 + (60 + k)s + 3K = 0$$

The R-H tabulation in the form of an array is shown below:

s^5	1	54	$60 + K$
s^4	13	82	$3K$

s^3	47.7	$60 + 0.769K$	0
s^2	$65.6 - 0.212K$	$3K$	0
s^1	$\dfrac{3940 - 105K - 0.163\,K^2}{65.6 - 0.212K}$	0	0
s^0	$3K$	0	0

For a stable system, the quantities of the first column in the R-H array should be greater than zero. The following equations must hold:

$$65.6 - 0.212K > 0 \text{ or } K < 309$$

$$3940 - 105K - 0.163K^2 > 0 \text{ or } K < 35$$

$$K > 0$$

For a stable system, $0 < K < 35$. Also, the value of K when the root loci cross the imaginary axis is 35. The radian frequency at the interception is determined from the auxiliary equation:

$$A(s) = (65.6 - 0.212K)s^2 + 3K = 0 \text{ (the fourth row of the R-H array)}$$

Substituting $K = 35$ into the auxiliary equation yields the following result:

$$A(s) = 58.2s^2 + 75 = 0$$

Solving for s in this equation yields:

$$s = \pm j1.13$$

This means radian frequency ω equals 1.13, which is the crossing point of the root locus diagram and is the frequency at which the system becomes unstable.

A breakaway point is between the two poles at -5 and -6, since the two loci which started from these poles meet and break toward infinity along the $+135°$ and the $-135°$ asymptotes, respectively. The breakaway point, $-a$, is readily determined from Equation 5-17:

$$\frac{-1}{6-a} = \frac{1}{a-3} - \frac{1}{a-5} - \frac{1}{a} - \frac{2(a-1)}{(a-1)^2+1}$$

Because this equation is of a high order, it is easier to solve for a by trial and error. Since it is known that a must be somewhere between 5 and 6, select as a first trial a = 5.5. The above equation then becomes:

$$\frac{-1}{0.5} = \frac{1}{2.5} - \frac{1}{0.5} - \frac{1}{5.5} - \frac{1}{21.25}$$

Solving for both sides of the above equation, you have $-2 = -2.205$, which is impossible. This means you must try another value for a, such that the left side of the equation becomes larger and the right side of the equation becomes smaller. Select a = 5.6.

$$\frac{-1}{0.4} = \frac{1}{2.52} - \frac{1}{0.6} - \frac{1}{5.6} - \frac{1}{22.2}$$

Solving for both sides of the equation, $-2.52 = -1.875$, which is again impossible. The next step is to choose a between 5.5 and 5.6. Select a = 5.52, for instance.

$$\frac{-1}{0.48} = \frac{1}{2.52} - \frac{1}{0.52} - \frac{1}{5.52} - \frac{1}{21.4}$$

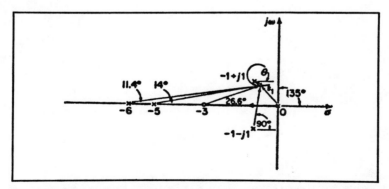

Fig. 5-17. Computation of the angle of departure of the root locus leaving the pole at $-1 + j1$.

135

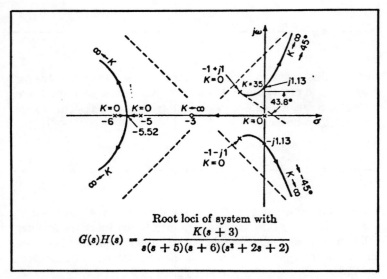

Root loci of system with

$$G(s)H(s) = \frac{K(s + 3)}{s(s + 5)(s + 6)(s^2 + 2s + 2)}$$

Fig. 5-18. Root loci of system with $G(s)H(s) = K(s + 3)/s(s + 5)(s + 6)(s^2 + 2s + 2)$.

Solving for both sides of the equation, $-2.08 = -2.13$, which is a close enough approximation. Therefore, $a = 5.52$ as the breakaway point. From the information obtained in these steps, the complete root locus diagram can be drawn, as shown in Fig. 5-18. The preceding example should be used as a model to construct most root loci plots of feedback control system loop transfer functions.

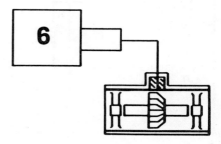

Control System Components

Control system components could be the proper subject of an entire book. There are numerous systems, such as thermal, mechanical, hydraulic pneumatic, electric, electronic, electro-mechanical, etc. Hence, there are as many types of components as there are types of physical systems. This chapter includes a number of control system components that represent a reasonable cross section of those components commonly employed in most control systems.

The description, characteristics, performance and mathematical model of each component will be considered on an individual basis. The examples presented for each of the components is only a means to an end, and the reader can employ his or her imagination to use these components in other configurations and systems.

The basic component areas that will be illustrated are transducers, electronic devices, error-sensing devices, and servo motors. Transducers include microphones, strain gauges, temperature sensing devices, and optical devices. The electronic devices include operational amplifiers and digital-to-analog converters. Error-sensing devices include synchros, tachometers, and accelerometers. Of course, both AC and DC servo motors will be included in this chapter.

OPERATIONAL AMPLIFIER CHARACTERISTICS AND CIRCUITS

The operational amplifier (op amp) is a linear integrated circuit (IC) which employs several differential amplifier stages in

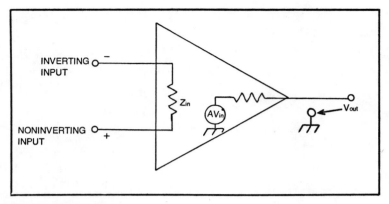

Fig. 6-1. Ideal op amp.

cascade. The differential amplifiers require both positive and negative power supplies; therefore, op amp circuits require two opposite polarity power supplies usually of equally equal value; i.e., (+15V and −15V). The ideal op amp, shown in Fig. 6-1, offers the following characteristics:

☐ Infinite input impedance Z_{in}.
☐ Zero output impedance Z_{out}.
☐ Infinite voltage amplification A_v.
☐ Infinite bandwidth.
☐ When the input voltage is zero, the output voltage is zero.
☐ Instantaneous recovery from saturation.

The *ideal op amp characteristics* means that almost any low-level signal can turn it on; there is practically no loading of one amplifier stage cascaded to another stage, and the op amp itself can drive an infinite number of other devices. However, the real world of electronics tells us that op amps have an input impedance of 100 MegΩ, an output impedance of 10Ω, and a voltage gain of 100,000, which drops sharply as frequency is increased.

Figure 6-2 is the schematic of a simple current amplifier. In the circuit, the output voltage is equal to the product of the output current and the feedback resistor, R_f. This result is accomplished through the fact that the op amp input impedance is much larger than R_f and R_{in} so that the op amp draws *zero* current. An expression for voltage gain A and input impedance Z_{in} can be developed.

At the input node to the (−) side of the amplifier we can write $I_{in} = I_f + I_a$, but I_a is zero because the op amp has a very high input impedance. Thus, $I_{in} = I_f$.

$$V_{in} = I_{in} R_{in} = I_f R_{in}$$

$$V_{out} = I_f R_f$$

$$A = \frac{V_{out}}{V_{in}} \qquad = \frac{-I_{fn} R_f}{I_f R_{in}}$$

$$A = \frac{-R_f}{R_{in}} \qquad \text{Equation 6-1}$$

$$Z_{in} = \frac{V_{in}}{I_{in}}$$

$$I_{in} = \frac{V_{out} - V_{in}}{R_f}$$

$$Z_{in} = \frac{V_{in} R_f}{V_{in} - V_{out}} \qquad = \frac{V_{in} R_f}{V_{in} - (-V_{in} A)}$$

$$Z_{in} = \frac{R_f}{1 + A} \qquad \text{Equation 6-2}$$

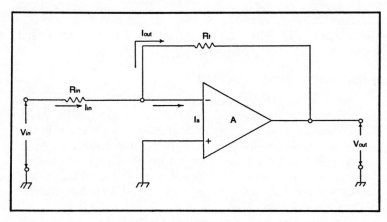

Fig. 6-2. Basic op amp amplifier circuit.

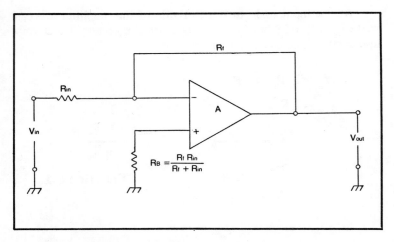

Fig. 6-3. Op amp amplifier circuit with biasing resistor.

For Equation 6-2, if the absolute magnitude of R_f approaches the absolute value of the gain of the device, this is no longer an op amp, and Equation 6-2 is no longer valid. For large values of A, Z'_{in} is so small that the input terminal marked (−) is called virtual ground. This input becomes a null point, which means the current from an input source will flow as though the source were returned to ground.

Figure 6-3 further develops the circuit of Fig. 6-2 by adding a bias resistor, R_b, to minimize the offset voltage at the output of the amplifier resulting from input-bias current. The value of the bias resistor is equal to the parallel combination of R_f and R_{in}. The circuit in Fig. 6-3 is the basic the basic op amp circuit used to calculate closed loop response.

FREQUENCY RESPONSE AND GAIN

Some of the design considerations for IC op amps are the result of trade-offs between gain and frequency response, which is also called *bandwidth*. The open loop (no feedback) gain and frequency responses are characteristics of the basic IC circuit, but these internal characteristics can be changed with external compensation networks. The closed loop (feedback) gain and frequency response are primarily dependent on the external feedback components.

The inverting and noninverting feedback op amp circuits appear in Fig. 6-4 and Fig. 6-5, respectively. Loop gain in these

140

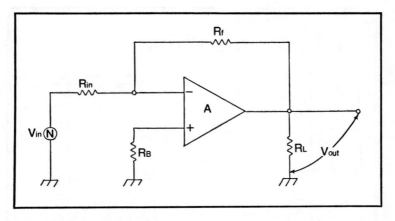

Fig. 6-4. Inverting feedback op amp.

figures is defined as the ratio of open-loop to close-loop gain, as shown in Fig. 6-6. The open-loop gain rolls off at the specified op amp characteristic, which is shown to be 6 dB/octave or 20 dB/decade. (The term 6 dB/octave means that the gain drops be 6 dB each time the frequency is doubled. Also, this is the same as a 20 dB drop in gain each time the frequency is increased by a factor of 10.)

If the open-loop gain of an amplifier was given in Fig. 6-6, any stable closed loop gain could be produced by the proper selection of feedback components, provided the closed loop gain

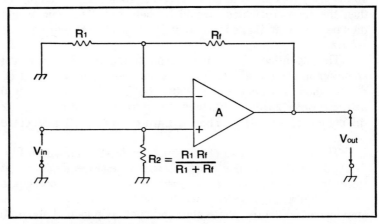

Fig. 6-5. Noninverting feedback op amp.

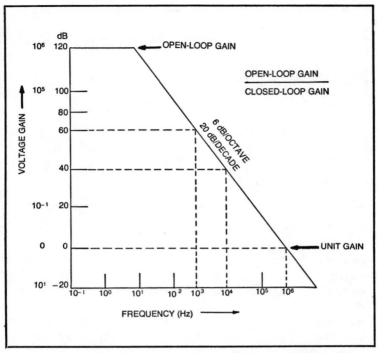

Fig. 6-6. Frequency response curve of a theoretical operational amplifier.

was less than the open loop gain. The main concern is a trade-off between gain and frequency response. For example, if a gain of 40 dB (10^2) was desired, a feedback resistance 10^2 times higher than the input resistance would be selected. The gain would then be flat to 10^4 Hz and roll off at 20 dB/decade to unity gain at 10^6 Hz.

The open-loop frequency response curve of a practical amplifier is shown in Fig. 6-7. The open-loop gain is flat at 60 dB to about a frequency of 200 kHz. Then it rolls off at 6 dB/octave to 2 MHz. Further on, a rolloff continues at 12 dB/detave to 20 MHz. Then a thirdroll off of 18 dB/octave occurs.

The phase response of the amplifier is also shown on Fig. 6-7. The phase response indicates that a negative feedback (at lower frequencies) can become positive and cause the op amp to be unstable at high frequencies (possibly oscillations). The op amp has a 180-degree phase shift and open-loop gain of about 20 dB.

In practice, when a selected closed-loop gain is equal to or less than the open-loop gain at the 180-degree phase point, the circuit will be unstable. For example, if a closed-loop gain of 20 dB or less had been selected, a circuit with the curves of Fig. 6-7 would be unstable. Therefore, the closed-loop gain must be more than the open-loop gain at the frequency where 180-degree phase shift occurs. The closed-loop gain would then have to be greater than 20 dB, but less than 60 dB.

OP AMP SPECIFICATIONS

The terms presented are most often used in op amp specifications as a measure of amplifier performance. an *input bias current*, I_B, is defined as the average of the two currents flowing into the input terminals of the op amp. In equation form, the input bias current can be defined as:

$$I_B = \frac{I_{B1} + I_{B2}}{2}$$ **Equation 6-3**

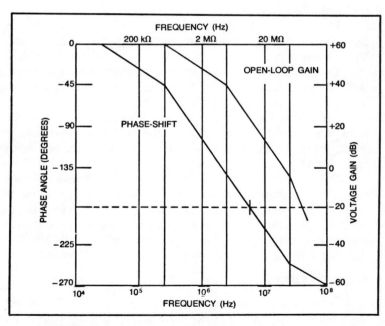

Fig. 6-7. Frequency response and phase-shift curve of a practical operational amplifier.

The base currents, I_{B1} and I_{B2}, are about equal to each other, which means that $I_B \cong I_{B1} \cong I_{B2}$. Depending on the type of op amp, the value of input bias current is usually small, generally in the range of 10 to 500 nA. Though this seems small, it can be a problem in circuits using relative large feedback resistors.

The *input offset current*, I_{os}, is defined as the difference between I_{B1} and I_{B2}.

$$I_{os} = I_{B1} - I_{B2} \qquad \textbf{Equation 6-4}$$

Typical values for input offset currents range from 0.05 to 150 nA.

Input offset voltage, V_{os}, is the voltage that must be applied between the input terminals through two equal resistances to force the output voltage to zero. When the output voltage is at zero, the op amp is said to be *nulled* or *balanced*. Typical values for input offset voltages range from 0.2 to 8 mV.

The op amp *input impedance*, Z_{in} is defined as the ratio of the change in input voltage to the change in input current. Typical values of input impedance range from 100 kΩ to 30 MΩ.

The op amp *output impedance*, Z_{out} is defined as the ratio of the change in output voltage to the change in output current. Typical values of output impedance ranges from 1 to 100 ohms.

The *common mode rejection ratio*, CMRR, is defined as the ratio of the input voltage range to the peak-to-peak change in input offset voltage over this range. Typcial values of CMRR range from 60 to 120 dB.

The *slew rate*, dv/dt, is the rate of output voltage change caused by a step input voltage and is usually measured in volts/microsecond. Typical values of slew rate range from 0.1 to 1000 V/μS. The slew rate specification applies to transient response. For a step function input, the slew rate tells how fast the output voltage can swing from one voltage level to another.

WHICH OP AMP DO YOU USE?

The designer must select an op amp that meets the requirements of the job. The specifications on the op amp (e.g., MC1741, see Appendix C) data sheet do not tell the whole story. Factors not readily obtainable from the data sheet which can affect the selection of an op amp include reliability, acceptability, capability to withstand environmental stresses, and, of course, cost. Decide which specifications are critical and which are not. For example, in building a voltage regulator with op amp ICs where there is a large differential voltage during turn-on, this must be considered to be a very critical parameter. In active filter design, a very high-gain, high-source impedance amplifier must be used; therefore, both offset current and offset voltage are critical parameters.

Besides selecting which specifications are needed and which are not, choices regarding safety factors might be required. Problems that will arise under fault conditions should be considered. For example, IC op amps can be bought with or without protection against output short circuit and input overvoltage. Because the emitter-to-base junctions of input transistors are sensitive to damage by large applied voltages, some form of input protection might be desirable. Ungrounded soldering irons, excessive input signals, and static discharges are all apt to challenge the input of the IC.

The op amp designer should know the roll off curve of his amplifier order to build a circuit with adequate gain stability over the working frequency range. The manufacturer may be perfectly justified in departing from the conventional 6 dB/octave frequency compensation to achieve such desirable features as fast settling time, high slew rate, fast overload recovery, or increased gain stability over a wide range of frequencies. To obtain these improved features, however, generally requires

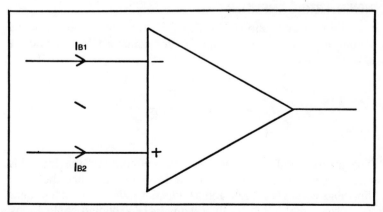

Fig. 6-8. Base bias current for an op amp.

fast rolloff characteristics and, therefore, a tendency toward oscillation.

Another often overlooked characteristic is the *slew rate*. Slew rate for the most part is just another way of looking at the rate limiting of the circuitry of the amplifier. The slew rate specification applies to transient response, whereas full power response applies to steady state of continuous response. For a step function input, slew rate tells how fast the output voltage can swing from one voltage level to another. Fast amplifiers will slew at up to 2000 V/μS, but amplifiers designed for DC applications often slew at 0.1V/μs.

INVERTING OP AMP

The basic inverting op amp circuit is shown in Fig. 6-9. The triangle that represents the IC op amp is assumed to depict the ideal op amp circuit. In other words, the open-loop amplifier gain is infinite (practically 200,000), the amplifier input impedance is infinite (actually 2 MΩ, and the amplifier output impedance is zero (actually 100 ohms).

The circuit shown in Fig. 6-9 gives a closed-loop gain of R_2/R_1. The input impedance is equal to R_1. The closed-loop bandwidth is equal to the unity gain frequency divided by 1, plus the closed-loop gain. The resistor, R_3 and R_2 to minimize the offset voltage error caused by bias current. Amplifier offset voltage is the predominant error for low source resistances, and offset current causes the main error for high source resistances.

NONINVERTING AMPLIFIER

The cirucit shown in Fig. 6-10 is of a noninverting amplifier. The output voltage of the noninverting amplifier is given by equation:

$$V_{out} = V_{in} \quad \frac{R_1 + R_2}{R_1}$$

The primary difference between this circuit and the inverting circuit are that the output is not inverted and that the input impedance is very high and is equal to the differential input impedance multiplied by loop gain. In DC-coupled applications,

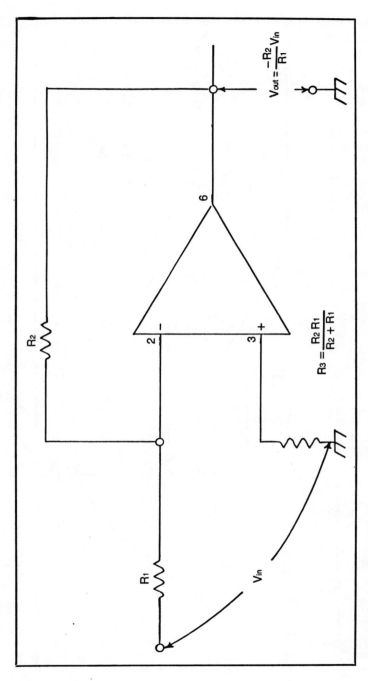

$$V_{out} = \frac{-R_2}{R_1} V_{in}$$

$$R_3 = \frac{R_2 R_1}{R_2 + R_1}$$

Fig. 6-9. Inverting amplifier.

147

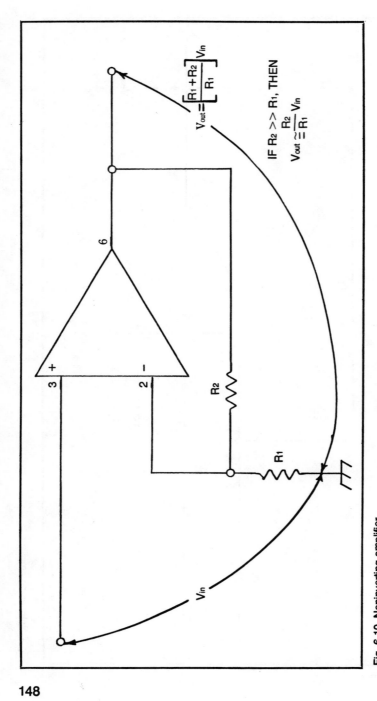

$$V_{out} = \left[\frac{R_1 + R_2}{R_1}\right] V_{in}$$

IF $R_2 >> R_1$, THEN

$$V_{out} \cong \frac{R_2}{R_1} V_{in}$$

Fig. 6-10. Noninverting amplifier.

input impedance is less important than input current and voltage drop across the course resistance.

In the equation for output voltage of the noninverting amplifier, if the resistance R_1 is much smaller than R_2, the output voltage can be written as follows:

$$V_{out} = V_{IN} \quad \frac{R_2}{R_1}$$

Application precautions are the same for this amplifier as for the inverting amplifier, with one exception. The amplifier output will go into saturation if the input is allowed to float. This may be important if the amplifier must be switched from source to source. The compensation trade-off discussed for the inverting amplifier is also valid for this connection.

VOLTAGE FOLLOWER

The voltage follower is frequently used as a buffer amplifier to reduce voltage error caused by source loading and to isolate high-impedance sources from following circuitry. Figure 6-11 shows a voltage follower op amp. The gain of the circuit is unity. The output follows the input voltage; hence, the name voltage follower.

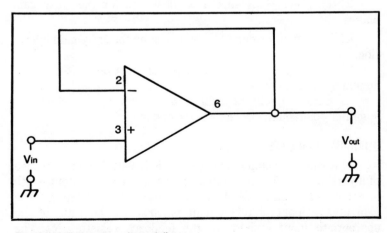

Fig. 6-11. Unity gain voltage follower.

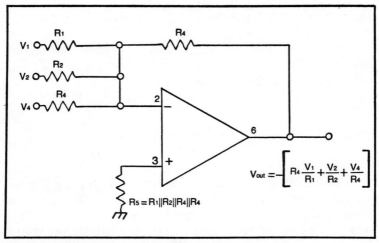

Fig. 6-12. Inverting summing amplifier.

The unity gain buffer provides the highest input impedance of any op amp circuit. Input impedance is equal to the differential input impedance multiplied by the open loop gain, in parallel with common-mode input impedance. The gain error of this circuit is equal to the reciprocal of the amplifier open-loop gain or to the common-mode rejection, whichever is less. Bias current for the amplifier is supplied by the source resistance and causes an error at the amplifier input because of its voltage drop across the source resistance.

Three precautions should be observed for the unity gain buffer:

☐ The amplifier must be compensated for unity gain operation.

☐ The output swing of the amplifier must be limited by the amplifier common mode range.

☐ Some amplifiers exhibit a latchup mode when the amplifier common mode range is exceeded.

SUMMING AMPLIFIER

The summing amplifier shown in Fig. 6-12 is a special case of inverting amplifier. The circuit gives an inverted output which is equal to the weighted algebraic sum of all three inputs. The gain of any input of this circuit is equal to the ratio of the appropriate input resistor to the feedback resistor, R_4. The

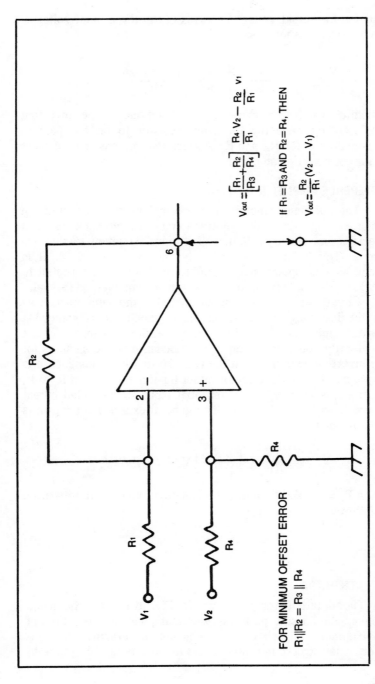

$$V_{out} = \left[\frac{R_1 + R_2}{R_3 + R_4} \right] \frac{R_4 \cdot V_2}{R_1} - \frac{R_2}{R_1} V_1$$

If $R_1 = R_3$ AND $R_2 = R_4$, THEN

$$V_{out} = \frac{R_2}{R_1} (V_2 - V_1)$$

FOR MINIMUM OFFSET ERROR

$R_1 \| R_2 = R_3 \| R_4$

Fig. 6-13. Difference amplifier.

151

equation for output voltage in terms of the three input voltages can be written as follows:

$$V_{out} = -R_4 \left[\frac{V_1}{R_1} + \frac{V_2}{R_2} + \frac{V_3}{R_3} \right]$$

Amplifier bandwidth may be calculated as in the inverting amplifier, by assuming the input resistor to be the parallel combination of R_1, R_2, and R_3. Application cautions are the same as for the inverting amplifier.

DIFFERENCE AMPLIFIER

The difference amplifier is the complement of the summing amplifier and allows the subtraction of two voltages or as a special case, the cancellation of a signal common to the two inputs. The difference amplifier is shown in Fig. 6-13. It is useful as a computational amplifier, in making a differential to single ended conversion, or in rejecting a common-mode signal.

Circuit bandwidth can be calculated in the same manner as for the inverting amplifier, but input impedance is somewhat more complicated. Input impedance for the two inputs is not necessarily equal; inverting input impedance is the same as for the inverting amplifier, and the noninverting input impedance is the sum of R_3 and R_4. Gain for either input is the ratio of R_1 to R_2 for the special case of a differential input single ended output where $R_1 = R_3$ and $R_2 = R_4$. The general expression for gain is the following:

$$V_{out} = \left[\frac{(R_1 + R_2) R_4}{(R_3 + R_4) R_1} \right] V_2 - \left(\frac{R_2}{R_1} \right) V_1$$

When $R_1 = R_3$ and $R_2 = R_4$, the equation for V_{out} can be written as follows:

$$V_{out} = \frac{R_2}{R_1} (V_2 - V_1)$$

DIFFERENTIATOR

The differentiator is shown in Fig. 6-14. As the name implies, the circuit performs the mathematical operation of *differentiation*. The circuit shown is not the practical differentiator. It is a true differentiator and is extremely susceptible to

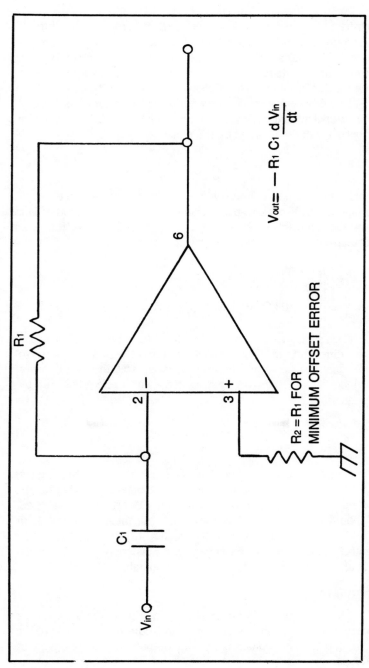

$$V_{out} = -R_1 C_1 \frac{d V_{in}}{dt}$$

$R_2 = R_1$ FOR MINIMUM OFFSET ERROR

Fig. 6-14. Differentiator.

153

high-frequency noise because the gain increases at the rate of 6dB/octave. In addition, the feedback network of the differentiator, R_2C_1, is an RC low-pass filter which contributes 90-degree phase shift to the loop and may cause stability problems even with an amplifier that is compensated for unity gain.

The differentiator circuit of Fig. 6-15 provides an output proportional to the derivative of the input signal. The equation for output voltage is given by the equation:

$$V_{out} = -R_2C_1 \frac{dV_1}{dt}$$

A triangular input voltage will produce a square wave output voltage. A 2.5V peak-to-peak triangle wave with a period of 1mS with the circuit shown in Fig. 6-15 develops the following calculations:

$$dV = \left\| \frac{2.5V}{0.5 \text{ mS}} = 5 \frac{V}{mS} \right.$$
$$V_{out} = \left\| -(10 \text{ k}\Omega)(0.1 \text{ }\mu\text{F})(5 \text{ V/mS}) = 5 \text{ V peak-to-peak} \right.$$

The resistor, R_1, is needed to limit the high-frequency gain of the differentiator. This makes the circuit less susceptible to high-freqeuncy noise and assures dynamic stability. The corner frequency where the gain limiting comes into effect is given by the equation:

$$f = \frac{1}{2\pi R_1 C_1}$$

The corner frequency should be at least 10 times the highest input frequency for accurate operation. A maximum value for the corner frequency is determined by stability criteria. In general, it should be no larger than the geometric mean between $1/2\pi$ R_2C_1 and the gain bandwidth product of the op amp. The differentiator is subject to damage from fast rising input signals. It is also susceptible to high-frequency instability.

INTEGRATOR

The integrator circuit is shown in Fig. 6-16. This circuit provides an output that is proportional to the time integral of the input signal. The equation for the output voltage is given as follows:

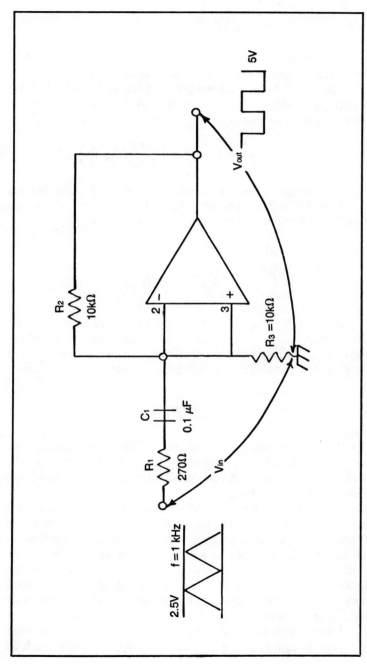

Fig. 6-15. Practical differentiator.

$$V_{out} = -\frac{1}{R_1C_1} \int V_{in} dt$$

As an example, consider the response of the integrator to a symmetrical square wave input signal with an average value of 0V. If the input has a peak amplitude of BV in the period T, the peak-to-peak output can be calculated by integrating over one-half the input period as illustrated by following equation:

$$|V_{out}| = \frac{1}{R_1C_1} \int_0^{T/2} B\, dt = \left(\frac{B}{R_1C_1}\right)\left(\frac{T}{2}\right) V$$

The wave shape will be triangular, corresponding to the integral of the square wave. For the component values shown in Fig. 6-16, B = 5V and T = 1 mS, the following results;

$$R_1C_1 = 10^{-3} \text{ S}$$

$$V_{out} = \left(\frac{5}{10^{-3}}\right)\left(\frac{10^{-3}}{2}\right) = 2.5V \text{ peak-to-peak}$$

Resistor R_2 is included to provide DC stabilization for the integrator. Its function is to limit the low-frequency gain of the amplifier and thus minimize drift. The frequency above which the circuit will perform as an integrator is given by the equation:

$$f = \frac{1}{2\pi R_2C_1}$$

For the best linearity, the frequency of the input signal should be at least 10 times the frequency given in the equation. The linearity of the circuit illustrated is better than 1 percent with an input frequency of 1 kHz.

Although it is not immediately obvious, the integrator, if it is to operate reliability, requires both a large common-mode and differential-mode input voltage range. There are several ways the input voltage limits may be inadvertently exceeded. The most obvious is that transients occurring at the output of the amplifier can be coupled back to the input by integrating capacitor C_1. Thus, either common-mode or differential-mode voltage limits can be exceeded.

Another less obvious problem can occur when the amplifier is driven from fast rising or fast falling input signals, such as

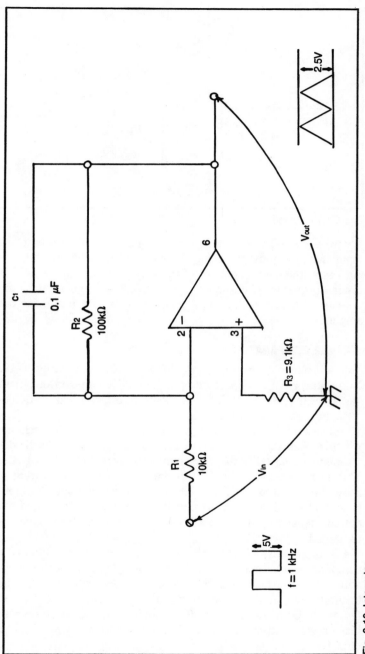

Fig. 6-16. Integrator.

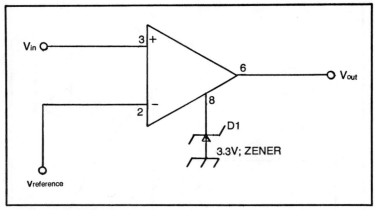

Fig. 6-17. Comparator circuit.

square waves. The output of the amplifier cannot respond to an input instantaneously. During the short interval before the output rests, the summing point of the amplifier may not be held at ground potential. If the input signal change is large enough, the voltage at the summing point could exceed safe limits for the amplifier.

VOLTAGE COMPARATOR

A voltage comparator amplifier is shown in Fig. 6-17. Notice the circuit is operated open loop. The comparator circuit has a variety of applications, including interface circuits, detectors, and sense amplifiers.

The circuit shown in Fig. 6-17 shows a clamping scheme which makes the output signal directly compatible with diode-transistor logic (DTL) or transistor-transistor logic (TTL) ICs. This is accomplished by the breakdown diode with a rating of 3.3V. The input at pin 2 is fixed to a reference voltage. All input signals at pin 3 are compared to the reference voltage, and the output doesn't respond until the reference voltage level is exceeded.

Figure 6-18 shows the connection of an op amp as a comparator and lamp driver. Transistor Q1 switches the lamp, with resistor R_2 limiting the current surge resulting from the turning on a cold lamp. Resistor R_1 determines the base drive to Q1, while D1 keeps the amplifier from putting excessive reverse bias on the emitter base junction of the lamp driver when it turns off.

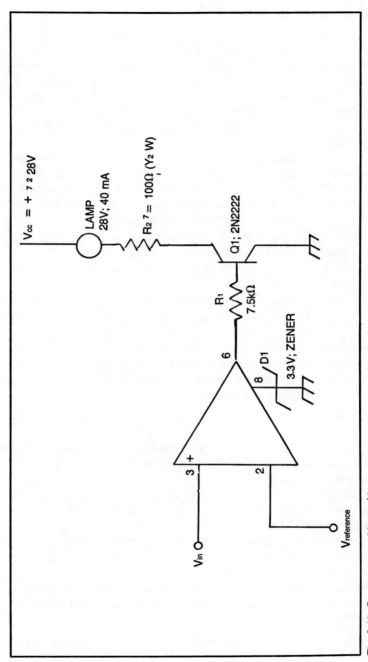

Fig. 6-18. Comparator and lamp drive.

159

DC SERVO MOTOR

The basic DC servo motor employed in a control system is an armature-controlled DC motor. This DC motor is used in velocity and position control systems to convert electric signals into a rotary or linear motion. They are extremely versatile drives, capable of reversible operation over a wide range of speeds, with accurate control of the speeds at all times. DC motors are available with horsepower ratings from 1/500 hp to over 500 hp.

The operating characteristics of electric motors and generators depend on the following basic fact: A force is exerted on an electric charge whenever it moves through a transverse magnetic field. This phenomenon is responsible for the following two actions which occur in motors and generators.

☐ A force is exerted on a conductor in a magnetic field whenever a current passes through the conductor.

☐ A voltage is induced in a conductor whenever it is moved through a magnetic field.

The force exerted on a current carrying conductor produces the torque that tends to rotate an electric motor. This force is also present in an electric generator whenever a current is passing through the armature winding. The direction of the force is perpendicular to both the direction of the current and the direction of the magnetic field. The magnitude of the force is given by the following equation:

$$F = iBL \sin\Theta \qquad \textbf{Equation 6-5}$$

where F = the force on the current carrying conductor in Newtons, i = the current through the conductor in amperes, L = the length of the conductor in meters, B = the flux density of the magnetic field that the conductor passes through in webers per square meter, and Θ = the angle between the direction of the current and the direction of the magnetic field in degrees.

The voltage induced in a moving conductor produces the voltage output of an electric generator. This induced voltage is also present in an electric motor whenever the armature is rotating. If the conductor and the magnetic field are mutually perpendicular, the induced voltage is given by the following equation.

$$v = LSB \sin\beta \qquad \textbf{Equation 6-6}$$

where v = The induced voltage in volts, L = The length of the conductor in meters, S = The velocity of the conductor in meters per second, B = The flux density in webers per square meter, and β = The angle between the direction of motion of the conductor and the direction of the magnetic field in degrees.

Figure 6-19 shows the fundamental construction of a DC motor. Four coils of wire are mounted in slots on a cylinder of magnetic material called the *armature*. The armature is mounted on bearings and is free to rotate in the magnetic field produced by the two field poles. The field poles may be permanent magnets or electromagnets, depending on the size of the motor (the smaller motors use permanent magnets). The ends of each coil are connected to adjacent segments of a segmented ring called the *commutator*. Electrical connection is made to the

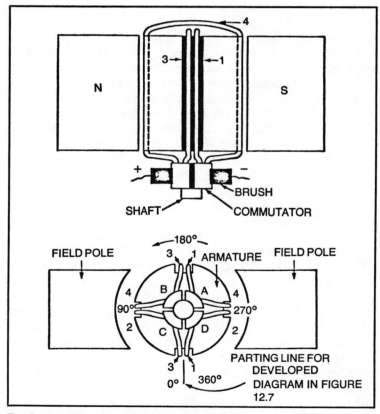

Fig. 6-19. Schematic diagram of a two-pole, four-coil DC motor.

armature coils through carbon contacts called *brushes*. Commercial motors have either two poles or four poles. In a four-pole motor, the opposite field poles are placed 90 degrees apart. The coil sides are placed one pole-span apart, or slightly more or less than one pole-span apart. The pole span is 180 degrees in a two-pole motor and 90 degrees in a four-pole motor.

The machine shown in Fig. 6-19 is capable of operating as either a DC generator or a DC motor. Its operation is based on these two facts:

☐ A voltage is induced in a conductor moving through a transverse magnetic field.

☐ A force is exerted on a current carrying conductor in a transverse magnetic field.

When the machine is used as a generator, the armature is rotated by an external prime mover usually mounted on a shaft placed through the armature. The motion of the coils through the magnetic field induces a voltage in the coil. The magnitude of the induced voltage is given by Equation 6-6. The polarity of the induced voltage reverses each time the coil passes from a north pole to a south pole, or vice versa. In a two-pole generator, the polarity reverses twice during each revolution. And in a four-pole generator, it reverses four times during each revolution. The function of the commutator is to reverse the connection between the brushes and the coil at the same time that the polarity of the induced voltage reverses. In other words, the commutator is a *rectifier* that converts an AC voltage into a DC voltage.

The generator then supplies current to a load connected to the armature conductors. As a result of this current, a force is exerted on the armature coils according to Equation 6-5. This force produces a torque that opposes the motion of the armature. The prime mover must overcome this additional torque to maintain the armature speed.

When the machine is used as a motor, a current is produced in the armature coils by connecting an external voltage source to the brushes. The current produces a force, according to Equation 6-5, which tends to rotate the armature. The commutator reverses the current in each coil as it passes from a north pole to a south pole, or vice versa. This reversal in the direction of the current eliminates the reversal in the direction of the force which would otherwise occur. The rotation of the armature also results in an induced voltage, according to Equation 6-6. This voltage opposes the external voltage and is called the back emf (electromotive force) of the motor.

A schematic diagram of the armature winding of the two-pole, four-coil motor is shown in Fig. 6-20. Notice in the figure

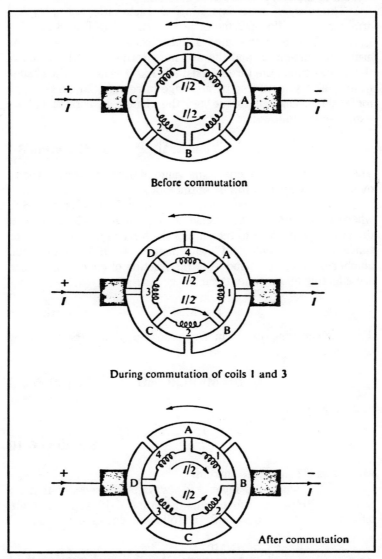

Before commutation

During commutation of coils 1 and 3

After commutation

Fig. 6-20. Schematic diagram of the armature winding of a two-pole, four coil DC motor.

that each coil carries one-half of the total armature current. As the armature rotates, both the polarity of the induced voltage and the direction of the current through the coil reverse each time a coil passes from one pole to the next. This reversal of current in the coil is called commutation.

The steady state operating characteristics of a typical armature-controlled DC motor are illustrated in Fig. 6-21. One graph in Fig. 6-21 indicates a linear relationship between the armature current, i, and the armature torque, T. The slope of this line is called the torque constant, K_T. It indicates the change in torque, ΔT, produced by a change in current, Δi. In equation form, the change in torque and the change in current can be expressed as follows:

$$\Delta T = K_T \Delta i \qquad \text{Equation 6-7}$$

The intercept on the i-axis is the value of the current required to overcome the friction torque of the motor.

The torque constant can be derived from Equation 6-5, where F is the force on a conductor of length L. The torque on the armature is equal to the sum of the forces on each conductor times the mean radius of the armature, R. If N is the total number of conductors and M is the fraction of conductors which are effective at any time, then:

$$T = MNFR \qquad \text{Equation 6-8}$$

Then substituting for F in Equation 6-5 yields the following equation:

$$T = MNBLRi \sin\Theta \qquad \text{Equation 6-9}$$

From Fig. 6-21:

$$i = \frac{i_a}{2} \qquad \text{Equation 6-10}$$

where i_a is the armature current.

In DC motors, the angle Θ is always 90 degrees, so that $\sin \Theta = \sin 90° = 1$. Substituting Equation 6-10 and the angle information into Equation 6-9, the following equation results:

$$T = \frac{(NMBLR)}{2} i_a \qquad \text{Equation 6-11}$$

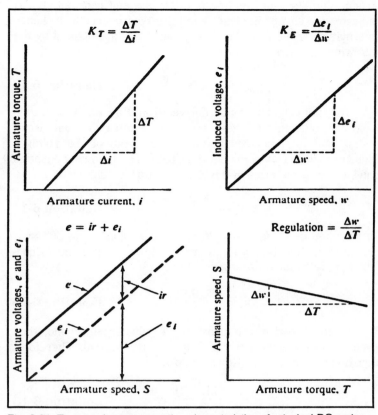

Fig. 6-21. The steady-state operating characteristics of a typical DC motor.

In this equation, the torque constant, K_T, is described by the following expression:

$$K_T = \frac{NMBLR}{2} \qquad \textbf{Equation 6-12}$$

For Equations 6-11 and 6-12, the following terms and units are defined. T = The torque in newton meters; K_T = The torque constant in newton-meters/ampere; i_a = The armature current in amperes; N = The total number of conductors; M = The fraction of effective conductors; B = The magnetic flux density in webers per square meter; L = The length of each conductor in meters; and R = The mean radius of the armature in meters.

From Fig. 6-21, a linear relationship between the armature speed, W, and the voltage induced in the armature coil, v_i, exists. The slope of this line is called the emf constant, K_E. It indicates the change in induced voltage Δv_i produced by a change in armature speed ΔW. These two quantities are defined by the following equation.

$$\Delta v_i = K_E \, \Delta W \qquad \textbf{Equation 6-13}$$

The emf constant can be derived from Equation 6-6. Notice from Fig. 6-20 that the coils are divided into two equal circuit paths. The total induced voltage is the sum of the voltages induced in each conductor in either path. The following equation can be written in terms of the total armature voltage, v_a.

$$v_i = \frac{NM}{2} \, v_a \qquad \textbf{Equation 6-14}$$

In Equation 6-6, angle $\beta = 90$ degrees so $\sin\beta = \sin 90° = 1$. Substituting Equation 6-14 and the angle information into Equation 6-6 yields the following equation:

$$v_1 = \frac{NMLSB}{2} \qquad \textbf{Equation 6-15}$$

If W is the speed of the armature in radians per second, then each conductor travels $S = RW$ meters per second. Therefore, equation 6-15 can be written as follows:

$$v_i = \frac{(NMBLR)}{2} \, W \qquad \textbf{Equation 6-16}$$

Thus, the emf constant can be expressed from Equation 6-16 as follows:

$$K_E = \frac{NMBLR}{2} \qquad \textbf{Equation 6-17}$$

Figure 6-21 shows a graph of the armature speed versus armature voltage with no load on the motor. The total armature voltage, v_a, is made up of two components: the induced voltage, v_i, and the iR drop across the armature resistance, R_a. Therefore, the total armature voltage is the sum of the two voltage components and can be expressed by the following equation:

$$v_a = iR_a + v_i \qquad \textbf{Equation 6-18}$$

Figure 6-21 shows a graph of the armature torque versus armature speed with a constant armature voltage. This graph is a result of the combined effects of Equation 6-7, 6-13, and 6-18. If the armature voltage v_a is constant, then according to Equation 6-18, any change in iR_a must be accompanied by an equal and opposite change in v_i. In other words,

$$\Delta v_i = - \Delta iR_a \qquad \text{Equation 6-19}$$

From Equation 6-7,

$$\Delta i = \Delta T/K_T \qquad \text{Equation 6-20}$$

From Equation 6-13,

$$\Delta v_i = K_E \Delta W \qquad \text{Equation 6-21}$$

Thus, from these equations,

$$\Delta v_i = K_E \Delta W = - \Delta iR_a = (-\Delta T) R_a/K_T \qquad \text{Equation 6-22}$$

or

$$K_E \Delta W = (-\Delta T) R/K_E \qquad \text{Equation 6-23}$$

From these equations the term *regulation* can be defined by the following equation:

$$\text{Regulation} = \frac{\Delta W}{\Delta T} = \frac{-R_a}{K_E K_T} \qquad \text{Equation 6-24}$$

Next, the important DC motor equations can be summarized:

$$v_a = iR_a + v_i \qquad \text{Equation 6-18}$$

$$v_i = K_E W \qquad \text{Equation 6-25}$$

$$T = K_T i - T_f \qquad \text{Equation 6-26}$$

$$REG = \frac{\Delta W}{\Delta T} = \frac{-R_a}{K_E K_T} \qquad \text{Equation 6-24}$$

$$P = WT \qquad \text{Equation 6-27}$$

167

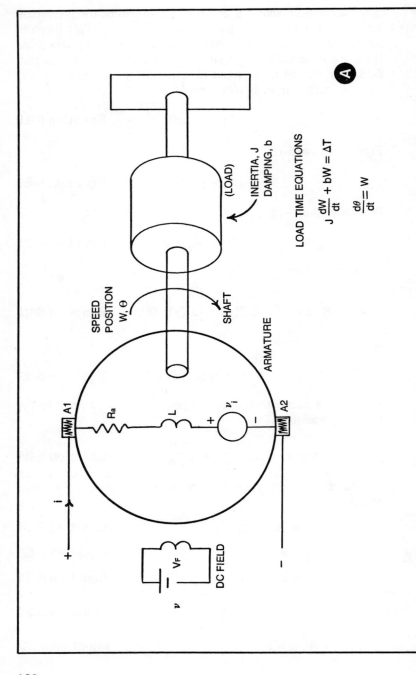

SPEED
POSITION
W, θ

SHAFT

ARMATURE

R_a

L

ν_i

A1

A2

i

DC FIELD

V_F

ν

(LOAD)

INERTIA, J
DAMPING, b

LOAD TIME EQUATIONS

$$J \frac{dW}{dt} + bW = \Delta T$$

$$\frac{d\theta}{dt} = W$$

A

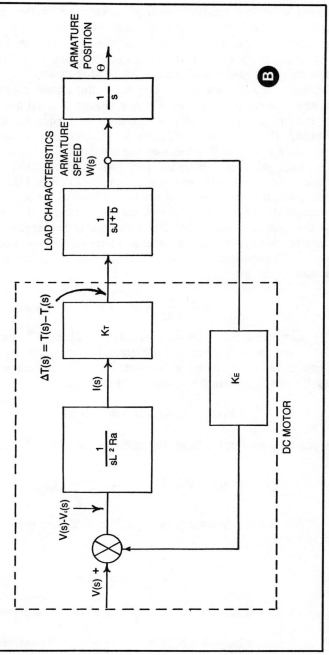

Fig. 6-22. Schematic of an armature-controlled DC motor connected to a load at A, and a block diagram of the system at B.

169

where v_a = the armature voltage in volts; v_i = the induced voltage involts; i_a = the armature current in amperes; R_a = the armature resistance, in ohms; K_E = the emf constant, in volts per radian per second; W = the armature speed, in radians per second; T = the output torque in newton-meters; T_f = the friction torque in newton-meters; K_T = the torque constant in newton-meters per ampere; ΔW = a change in armature speed in radians per second; ΔT = a change in torque in newton-meters; REG = the regulation in radians per second per Newton-meter; and P = the power in watts.

The schematic diagram and the block diagram of an armature-controlled DC motor are illustrated in Fig. 6-22. A DC voltage is applied to the field winding or the field is provided by permanent magnet field poles. A variable voltage (v) is applied to the armature windings. The armature is represented by a resistor, an inductor, and an induced voltage source connected in series. The armature current (i) is defined by the following time domain equation:

$$v = L\ \frac{di}{dt}\ + iR_a + v_1 \qquad \textbf{Equation 6-28}$$

Employing the methods illustrated in Chapter 3, Equation 6-28, can be transformed from time to frequency by using Laplace transforms listed in Appendix B and C. The corresponding frequency domain equation is written as follows:

$$V(s) = sLI(s) + I(s)\ R_a + V_i(s)$$

Solving for current I(s) in the equation yields:

$$I(s) = \left[V(s) - V_i(s)\right]\ \frac{1}{sL + R_a} \qquad \textbf{Equation 6-29}$$

The frequency domain of Equations 6-25 and 6-26 can be expressed:

$$T(s) \quad = \quad K_T I(s) - T_f(s)$$

$$K_T I(s) \quad = \quad T(s) - T_f(s) = \Delta T(s) \quad \textbf{Equation 6-30}$$

$$V_i(s) \quad = \quad K_E W(s) \qquad\qquad \textbf{Equation 6-31}$$

The load time equations are given in Fig. 6-21, and they can also be expressed in the frequency domain by the following equations:

$$\Theta(s) = W(s) \frac{1}{s} \qquad \text{Equation 6-32}$$

$$W(s) = \Delta T(s) \frac{1}{sJ + b} \qquad \text{Equation 6-33}$$

Remeber that in these equations, $\Delta T = T - T_f$.

These equations are each represented on the block diagram of Fig. 6-22B by employing the methods illustrated in Chapter 2. The important transfer functions of the DC motor armature control follow.

Velocity Transfer Function

$$\frac{W(s)}{V(s)} = \frac{K_T/R_a b}{T_m T_a s^2 + (T_m + T_a)s + \dfrac{(1 + K_E K_T)}{R_a b}} \qquad \text{Equation 6-34}$$

Position Transfer Function

$$\frac{\Theta(s)}{V(s)} = \frac{W(s)}{V(s)} \frac{1}{s} \qquad \text{Equation 6-35}$$

where $V(s)$ = The armature voltage in volts; $W(s)$ = The armature speed in radians per second; $\Theta(s)$ = The armature position in radians; $T_m = J/b$ in seconds; $T_a = L/R_a$ in seconds; J = The moment of inertia of the load connected to the shaft from the armature of the motor in kilogram square meters; b = The damping resistance of the load in newton-meters per radian per second; L = The armature inductance in henrys; R_a = The armature resistance in ohms; K_T = The torque constant, in newton-meters per ampere; K_E = the emf constant, in volts per radian per second; and s = The Laplace transform operator.

☐ **Example 6-1:** An armature-controlled DC motor has the following ratings:

$$T_f = 1.2 (10)^{-2} \text{ newton-meters}$$

$$K_T = 0.06 \text{ newton-meter per ampere}$$

$$I_{MAX} = 2 \text{ amperes}$$

$$K_E = 0.06 \text{ volts per radian per second}$$

$$W_{MAX} = 500 \text{ radians per second}$$

$$R_a = 1.2 \text{ ohms}$$

Find the maximum output torque, the maximum power output, the regulation, and the maximum armature voltage.

The maximum output is found from Equation 6-26 when $i = I_{MAX}$.

$$T_{MAX} = K_T I_{MAX} - T_f = (0.06)(2) - 0.012$$

$$= 0.108 \text{ newton-meters}$$

The maximum output power is found from Equation 6-27 when W and T are both maximum.

$$P_{MAX} = W_{MAX} T_{MAX} = (500)(0.108)$$

$$= 54 \text{ watts}$$

The regulation is found from Equation 6-24.

$$REG = \frac{-R_a}{K_E K_T} = \frac{-1.2}{(0.06)^2}$$

$$= -333 \text{ radians per second per newton-meter}$$

The maximum armature voltage is found from Equations 6-18 and 6-25, when i and W are both at maximum.

$$v_i = K_E W = (0.06)(500) = 30 \text{ volts}$$

$$V_{MAX} = I_{MAX} R_a + v_i$$

$$= (2)(1.2) + 30V$$

$$= 32.4V$$

□ **Example 6-2:** The motor in Example 6-1 is operated at 300 radians per second with a load torque of 0.05 newton-meters. Determine the armature voltage and the armature speed if the torque increases to 0.075 newton-meters and the armature voltage is not changed.

From Equation 6-26, the armature current is found:

$$i = \frac{T - T_f}{K_T} = \frac{0.05 + 0.012}{0.06}$$

$$= 1.03 \text{ amperes}$$

From Equations 6-18 and 6-25, the armature voltage is found:

$$v_i = K_E W = (0.06)(300) = 18V$$

$$v = iR_a + v_i = (1.03)(1.2) + 18V$$

$$= 19.24 \text{ Volts}$$

The regulation is determined from Equation 6-1, and from the regulation we can find the change in torque and then the armature speed.

$$\Delta W = (REG)(\Delta T)$$

$$\Delta T = 0.075 - 0.05 = 0.025$$

$$\Delta W = (-333)(0.025) = -8.33 \text{ radians per second}$$

$$W = 300 - 8.33 = 291.67 \text{ radians per second}$$

□ **Example 6-3:** Determine the velocity and positon transfer functions of the motor in Example 6-1. The following additional values are required.

$$J = 6.2 (10)^{-4} \text{ kilogram-square meter}$$

$$b = 1 (10)^{-4} \text{ newton-meter per radian per second}$$

$$L = 0.02 \text{ henry}$$

The following system constants are found.

$$T_m \quad = \quad J/b = 6.2\,(10)^{-4}/(10)^{-4} = 6.2 \text{ seconds}$$

$$T_a \quad = \quad L/R_a = 0.02/1.2 = 0.0167 \text{ seconds}$$

$$T_m T_a \quad = \quad (6.2)(0.0167) = 0.103$$

$$T_m + T_a \quad = \quad 6.2 + 0.0167 = 6.22 \text{ seconds}$$

$$K_T/R_a B \quad = \quad 0.06/(1.2)(10)^{-4} = 500$$

$$1 + \frac{K_E K_T}{R_a B} \quad = \quad 1 + 500(0.06) = 1 + 30 = 31$$

Thus the velocity transfer function is written as follows:

$$\frac{W(s)}{V(s)} = \frac{500}{0.103s^2 + 6.22s + 1}$$

The position control transfer function is written as follows:

$$\frac{\Theta(s)}{V(s)} = \frac{16.1}{s(3.32 \times 10^{-3}\,s^2 + 0.201\,s + 1}$$

TWO-PHASE AC MOTOR

Two-phase AC motors are often used in control systems which required a low-power, variable-speed drive. The primary advantage of the AC motor over the DC motor is its ability to use the AC output of synchros and other AC measuring means without demodulation of the error signal. An AC amplifier provides the gain for a proportional control mode. However, more elaborate control modes are difficult to implement with an AC signal. When additional control actions are required, the AC signal is usually demodulated, and the control action is inserted in the DC signal. The modified DC signal is then reconverted.

The schematic diagram of a two-phase AC motor is shown in Fig. 6-23. The motor consists of an induction rotor, and two field coils located 90 degrees apart. One field coil serves as a fixed reference field, the other as the control field. The amplified AC error signal is applied to the control field. This signal has a variable

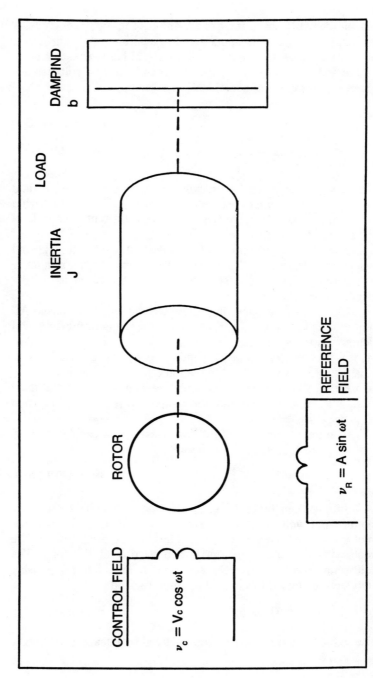

Fig. 6-23. A two-phase AC motor.

175

magnitude with a phase angle of either 0 degrees or 180 degrees. A constant AC voltage is applied to the reference field through a 90-degree phase shift network. This signal has a constant magnitude and a phase angle of -90 degrees. The two voltages are given in the following equations.

$$v_c = V_c \cos \omega t \qquad \text{Equation 6-36}$$

$$v_r = A \cos (\omega t - 90°) = A \sin \omega t \qquad \text{Equation 6-37}$$

where v_c = the control field voltage, v_r = the reference field voltage, ω = the operating radian frequency, V_c = the variable amplitude of the control votlage, A = the constant amplitude of the reference voltage.

The sign and magnitude of V_c is determined by the sign and magnitude of the error signal. A negative error signal results in a negative value of V_c. This is usually interpreted as a 180-degree phase shift in v_c.

The linearized operating characteristics of a two-phase AC motor are shown in Fig. 6-24. The actual operating line will depend on the speed torque characteristics of the process. Two typical process load lines are indicated by the dotted lines in Fig. 6-24. A change from one load line to another is an example of a process load change. The negative values of V_c in the third quadrant simply indicate that the motor reverses direction when V_c is negative.

The torque equation of the two-phase motor can be obtained from Fig. 6-24, and is the following:

$$T = K_1 V_c - K_2 W \qquad \text{Equation 6-38}$$

In Equation 6-38, K_1 is a constant with units of newton-meters per volt, and K_2 is a constant with units of newton-meters per radian per second. The torque produced by the motor is applied to the inertia and friction of the load. The following frequency domain equation defines the relationship between the motor torque and the motor speed.

$$T = (Js + b)W \qquad \text{Equation 6-39}$$

The velocity transfer function is obtained by equating Equations 6-38 and 6-39.

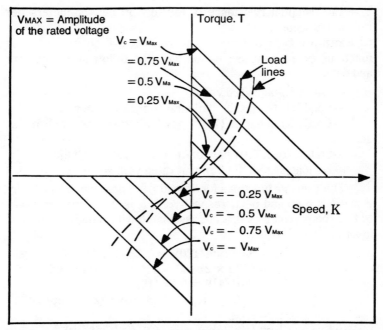

Fig. 6-24. The linearized operating characteristics of a two-phase AC motor.

A list of transfer functions and system constants for the AC two phase motor follow.

Velocity Transfer Function

$$\frac{W(s)}{V_c(s)} = \frac{K}{sT_t + 1} \qquad \textbf{Equation 6-40}$$

Position Transfer Function

$$\frac{\Theta(s)}{V_c(s)} = \frac{K}{s(sT_t + 1)} \qquad \textbf{Equation 6-41}$$

where V_c = the control voltage amplitude in volts, W = the motor speed in radians/second, Θ = the motor position in radians, $K = K_1/(b + K_2)$ **(Equation 6-42)**, $T_t = J/(b + K_2)$ in seconds **(Equation 6-43)**, K_1 = stall torque per rated voltage in newton-meters per volt, K_2 = stall torque per no load speed at the rated voltage in newton-meters per radian per second, J = the moment of inertia of the load, in kilogram-square meter, and

b = the damping resistance of the load in newton-meters per radian per second.

□ **Example 6-4:** Determine the velocity and position transfer functions of a two-phase motor with the following parameters and data:

—Rated voltage: 120 volts

—Load Inertia: 6 $(10)^{-6}$ kilograms-square meter

—Load damping: 2 $(10)^{-5}$ newton-meters per radian per second

—Stall torque: 0.04 newton-meters at rated voltage

—No load speed: 4000 rpm at rated voltage

The transfer functions for velocity and position are given by Equations 6-40 and 6-41, respectively. It is therefore necessary to find the parameters that are associated with the given transfer functions.

$$K_1 = 0.04 \text{ rpm}/120V = 3.33 (10)^{-4}$$
$$\text{No load speed} = 4000 \times 2\pi/60 = 419 \text{ radians/second}$$
$$K_2 = 0.04/419 = 9.56 (10)^{-5}$$
$$K = K_1/(b + K_2) = 3.33 (10)^{-4}/(2 + 9.56) (10)^{-5}$$
$$K = 2.88$$
$$T_t = J/(b + K_2) = 6 (10)^{-7}/(11.56) (10)^{-5}$$

The velocity transfer function is expressed by the following equation:

$$\frac{W(s)}{V_c(s)} = \frac{2.88}{0.052s + 1}$$

The position transfer function is expressed by the equation shown below.

$$\frac{\Theta(s)}{V_c(s)} = \frac{2.88}{s(0.052s + 1)}$$

ERROR-SENSING DEVICES

An error-sensing device compares two signals simultaneously. In feedback control systems, the purpose of an error-sensing device is to produce a signal that is proportional to the difference between the reference input and the controlled output variable. Linear wirewound potentiometers, properly connected,

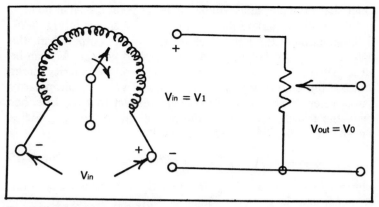

Fig. 6-25. Single-turn potentiometer.

are frequently used as error detectors. Other error detectors are control transformers and synchros.

Figure 6-25 shows a single-turn, wirewound potentiometer. The schematic diagram of the potentiometer is shown in Fig. 6-25B. If the potentiometer is assumed to be linear, the transfer function relating the output voltage (V_o) and the angular shaft rotation Θ_r is given by the following equation.

$$\frac{V_o(s)}{\Theta_r(s)} = \frac{V_i}{\Theta_{max}} \qquad \textbf{Equation 6-44}$$

Equation 6-44 implies that a single potentiometer can be used as an error detector simply by connecting the reference input shaft to the potentiometer case and the controlled shaft to the shaft of the potentiometer. However, this scheme is not commonly feasible because in feedback control systems, the controlled shaft and the reference input shaft are almost always remotely located. A practical application is to connect two potentiometers as shown in Fig. 6-26.

In Fig. 6-26, the two potentiometers convert the input and output shaft positions into proportional electric signals. The two electric signals are, in turn, compared and the difference voltage (v_e) is produced at the two terminals, a and b. From the electrical viewpoint, the schematic shown in Fig. 6-26 represents a simple bridge circuit. When the two arms are in the same relative positions, the potential difference between points a and b is zero. If the position of the output shaft is above the input shaft, the potential will be higher at point b than at point a, and the polarity mark of the error (v_e) is that shown in Fig. 6-26. If the position of the input is above that of the output shaft, the polarity of v_e will be reversed. The applied voltage, V_i, can be either DC or AC. If a DC voltage is applied, the polarity marks indicate the sign of v_e; for an AC applied voltage, the polarity marks refer to the phase of $v_i(t)$ with respect to $v_e(t)$. In either case, the transfer function of the potentiometer sensing device shown in Fig. 6-26 is written as follows:

$$v_e = K_s(\Theta_r - \Theta_c) \qquad \textbf{Equation 6-45}$$

where v_e = the error voltage in volts, K_s = the sensitivity of the error detector in volts per radian or volts per degree, Θ_r = the reference input shaft position in radians or degrees, and Θ_c = the controlled output shaft position in radians or degrees.

A typical application of the pair of potentiometers as an error detector in a servomechanism is shown in Fig. 6-27. In this system, an electric signal proportional to the misalignment between the reference input shaft and the controlled shaft is generated at the output of the potentiometer. The amplifier amplifies the electric signal until it is of sufficient power to drive the DC motor. The motor will then rotate in such a direction as to reduce error voltage v_e. Theoretically, when the error voltage is zero, the output shaft is in correspondence with the reference input shaft.

Most of the potentiometers used in servo applications are of the wirewound type. As the potentiometer shaft is rotated, the sliding brush contacts only discrete points on the wire. Thus the output voltage of the potentiometer is not an exact continuous function of the shaft rotation. The actual relation between v_o and Θ is shown in Fig. 6-28.

The resolution of a potentiometer is defined as the minimum change in output voltage V_o obtained by rotating the shaft, expressed as a percentage of the total applied voltage, V_i. If we let the number of turns of resistance wire be N, we can write the resolution in the following equation:

$$\text{Resolution} = \frac{\Delta V_o}{V_i} = \frac{V_i/N}{V_i} = \frac{1}{N} \qquad \textbf{Equation 6-46}$$

The resolution of a potentiometer places an upper limit on the accuracy of a servo system. The resolution of precision

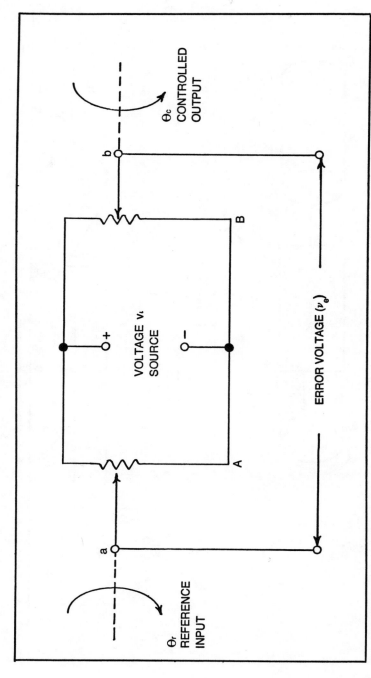

Fig. 6-26. Potentiometer sensing device.

181

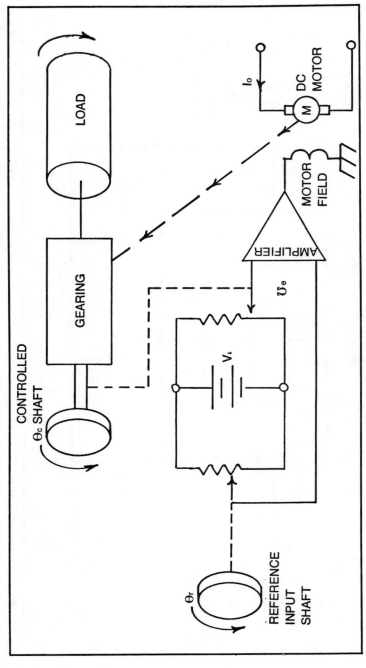

Fig. 6-27. Simple servomechanisms with potentiometers as a sensing device.

wirewound potentiometers ranges between 0.001 percent to 0.5 percent. For a carbon potentiometer, the resolution is zero, because the slider moves along a continuous resistance path.

Although the principle of operation and construction of a potentiometer is quite simple, its application in feedback control system is seldom satisfactory for the following major reasons:

☐ Precision wirewound potentiometers are very delicate devices. The sliding contact may be subject to damage if it is not handled carefully.

☐ The output voltage is discontinuous; thus it contributes to servo inaccuracy.

☐ For a single-turn potentiometer, the usable angle of rotation is less than 360 degrees.

☐ Because of the limit on heat dissipation of the potentiometer, the voltage applied to the potentiometer cannot be too large. The sensitivity, therefore, seldom exceeds 0.1V per degree; a higher gain amplifier is required than if synchros are used.

Among the various types of error-sensing devices, perhaps the most widely used in feedback control systems is a pair of synchros. Basically, a synchro is a rotary device used to produce a correlation between an angular position and a voltage or set of voltages. Depending upon the manufacturers, synchros are identified by such trade names as *Selsyn, Autosyn, Diehlsyn,* and *Telesyn.* There are several different types and applications of synchros, but only the synchro transmitter, the synchro control

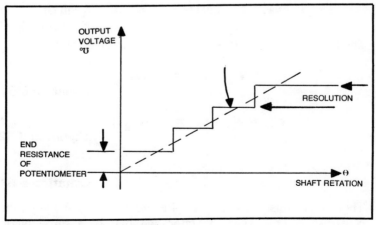

Fig. 6-28. Potentiometer resolution.

transformer, and the synchro differential transmitter will be discussed in this section.

A *synchro transmitter* has a Y-connected stator winding that resembles the stator of a three-phase induction motor. The rotor is a salient-pole, dumbbell-shaped magnet with a single winding. A single-phase excitation voltage is applied to the rotor through two slip rings. The voltage may be 115V at 60 hertz, 115V at 400 hertz, or some other voltage and frequency depending upon the rating of the synchro. Schematic diagrams of a synchro transmitter are shown in Fig. 6-29. The symbol, *G*, is often used to designate a synchro transmitter, which is sometimes known as a synchro *generator*.

Let the AC voltage applied to the rotor be as follows:

$$v_r(t) = V_R \sin \omega t$$

Equation 6-47

When the rotor is in the position shown in Fig. 6-29B, the voltage induced across stator winding S_2 and the neutral is maximum and is written as follows:

$$v_{S2n}(t) = KV_R \sin \omega t$$

Equation 6-48

where K is a proportional constant. The voltages appearing across terminals S_1n and S_2n are given by the following equations:

$$v_{S1n}(t) = KV_R \cos 240° (\sin \omega t) = -0.5KV_R \sin \omega t$$

Equation 6-49

$$v_{S2n}(t) = KV_R \cos 120° (\sin \omega t) = -0.5KV_R \sin \omega t$$

Equation 6-50

The three terminal voltages of the stator are given below.

$$v_{S1S2} = v_{S1n} - v_{S2n} = -1.5 KV_R \sin \omega t$$

Equation 6-51

$$v_{S2S3} = v_{S2n} - v_{S3n} = 1.5 KV_R \sin \omega t$$

Equation 6-52

$$v_{S3S1} = v_{S3n} - v_{S1n} = 0$$

Equation 6-53

These equations show that despite the similarity between the construction of a synchro stator and that of a three phase

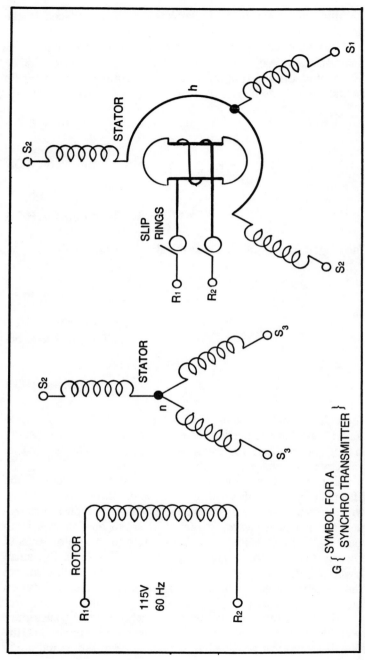

Fig. 6-29. Diagrams of a synchro transmitter.

185

machine, there are only single-phase voltages induced in the stator.

Consider now that the rotor is allowed to rotate in a counterclockwise direction. The voltages in each stator winding will vary as the function of the cosine of rotor displacement Θ; that is, the voltage magnitudes are stated as follows:

$$V_{S1n} = KV_R \cos (\Theta - 240°) \quad \textbf{Equation 6-54}$$

$$V_{S2n} = KV_R \cos \Theta \quad \textbf{Equation 6-55}$$

$$V_{S3n} = KV_R \cos (\Theta - 120°) \quad \textbf{Equation 6-56}$$

The magnitudes of the stator terminal voltages become:

$$V_{S1S2} = V_{S1n} + V_{nS2} = \sqrt{3} \, KV_R \sin (\Theta + 240°)$$
$$\textbf{Equation 6-57}$$

$$V_{S2S3} = V_{S2n} + V_{nS3} = \sqrt{3} \, KV_R \sin (\Theta + 120°)$$
$$\textbf{Equation 6-58}$$

$$V_{S3S1} = V_{S3n} + V_{nS1} = \sqrt{3} \, KV_R \sin \Theta$$
$$\textbf{Equation 6-59}$$

The voltage shaft position relationships specified in these equations imply that the synchro transmitter can be used to identify an angular position by measuring and identifying the set of voltages at the stator terminals.

Because the function of an error detector is to convert the difference of two shaft positions into an electric signal, a single synchro transmitter is apparently inadequate. A typical arrangement of a synchro error detector in servo applications is to connect the stator leads of the transmitter to the stator leads of a synchro control transformer, as shown in Fig. 6-30. For small deviations, the voltage at the rotor terminals of the control transformer in Fig. 6-30 is proportional to the deviation between the two rotor positions.

Basically, the construction of a synchro control transformer is similar to that of the synchro transmitter, except that the rotor is cylindrically shaped so that the air gap flux is uniformly distributed around the rotor. This is essential to a control transformer since its rotor terminals are usually connected to an amplifier; the change in the rotor impedance with rotation of the shaft should be minimized. The symbol, *CT*, is often used to designate a synchro control transformer.

The voltages given by Equations 6-57 through 6-59 are now impressed across the corresponding stator leads of the control transformer. Because of the similarity in the magnetic construction, the flux patterns produced in the two synchros will be the same if all losses are neglected. For example, if the rotor of the transmitter is in its electric zero position, the fluxes produced in the transmitter and in the control transformer are all vertical, as shown in Fig. 6-31.

When the rotor of the control transformer is in the position shown in Fig. 6-31B, the induced voltage at its rotor winding terminals is zero. The shafts of the two synchros are considered to be in alignment. However, when the rotor position of the control transformer is rotated 180 degrees from the position shown in Fig. 6-31B, the terminal voltage is again zero. These are known as the two *null positions* of the control transformer. If the control transformer rotor is rotated an angle α from either of the null positions, as shown in Fig. 6-31C, the magnitude of the rotor voltage is proportional to $\sin\alpha$. Similarly, it can be shown that when the transmitter shaft is in any position other than that shown in Fig. 6-31A, the flux pattern will change accordingly, and the rotor voltage of the control transformer will be proportional to the sine of the difference of rotor positions α. The rotor voltage of a control transformer versus the difference in positions of the rotors of the transmitter and control transformer is given in Fig. 6-32. For small angular displacement (about 15 degrees) in the vicinity of the two null positions, the rotor voltage of the control transformer is approximately proportional to the difference between the position of the rotors of the transmitter and the control transformer.

A typical AC servo system employing a synchro error detector is shown in Fig. 6-33. The purpose is to make the controlled shaft follow the angular variations of the reference input shaft. The rotor of the control transformer is mechanically connected to the controlled shaft. The rotor of the synchro transmitter is connected to the reference input shaft. The error signal which appears at the rotor terminals of the control transformer is amplified, and the amplified signal eventually

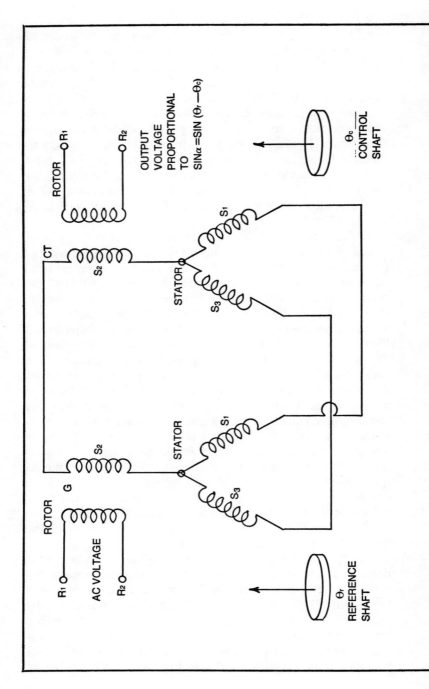

ROTOR R₁ R₂

OUTPUT VOLTAGE PROPORTIONAL TO $\sin \alpha = \sin(\theta_r - \theta_c)$

θ_c CONTROL SHAFT

CT S₂

STATOR S₁ S₃

STATOR S₁ S₃

G S₂

ROTOR AC VOLTAGE R₁ R₂

θ_r REFERENCE SHAFT

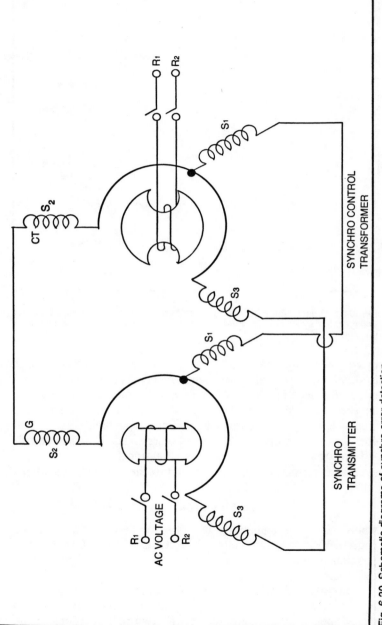

Fig. 6-30. Schematic diagrams of synchro error detection.

189

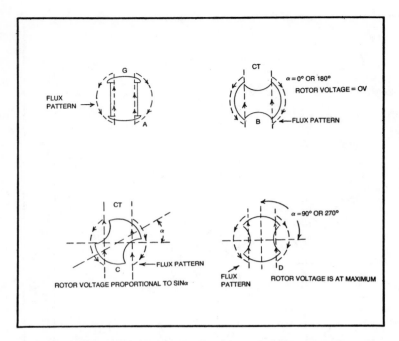

Fig. 6-31. Relationships between flux patterns, rotor positions, and the rotor voltage of a synchro error detector. The various positions are shown in A through D.

drives the two-phase induction motor. When the controlled shaft is aligned with the reference shaft, the error voltage is zero, and the motor does not turn. When an angular misalignment exists, an error voltage of relative polarity appears at the amplifier input. The motor will turn in the corresponding direction to correct this discrepancy. For small discrepancies of the controlled and reference shaft positions, the transfer function of the synchro error detector can be written by the following equation:

$$K_s = \frac{V_e}{\Theta_r - \Theta_c} \qquad \textbf{Equation 6-60}$$

where K_s = the sensitivity of the error detector in volts per degree, V_e = the error voltage in volts, Θ_r = the reference shaft position in degrees, and Θ_c = the controlled shaft position in degrees.

From the characteristics shown in Fig. 6-32, it is clear that K_s has opposite signs at the two null positions. However, in closed-loop systems, only one of the two null positions is the *true null*; the other one corresponds to an unstable operating point. Suppose that in the system shown in Fig. 6-33, the synchros are operating close to the true null, and the controlled shaft lags behind the reference shaft; a positive error voltage will cause the motor to turn in the proper direction to correct the lag. If the synchros are operating close to the false null, however, for the same lag between Θ_r and Θ_c, the error voltage is negative, which will cause the motor to run in the direction to increase the lag. A larger lag in the controlled shaft will increase the magnitude of the error voltage still further and cause the motor to turn further in the same direction until K_s is reversed in sign and the true null position is reached.

In practice, the discrepancy between the controlled shaft and the reference shaft may be represented as a function of time $\Theta_e(t)$. The error voltage is a modulated form as shown by the following equation:

$$v_e(t) = K_s\Theta_e(t) \sin \omega_c t \qquad \textbf{Equation 6-61}$$

In Equation 6-61, carrier frequency ω_c is the same as the frequency of the AC supply voltage. The magnitude of the modulated carrier wave is proportional to the shaft discrepancy, $\Theta_e(t)$, and the instantaneous polarity depends on the sign of the error.

Both the stator and the rotor of a synchro differential transmitter have distributed windings similar to the winding on

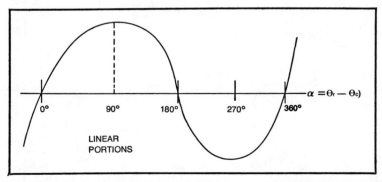

Fig. 6-32. Rotor voltage of control transformer as a function of the difference of rotor position.

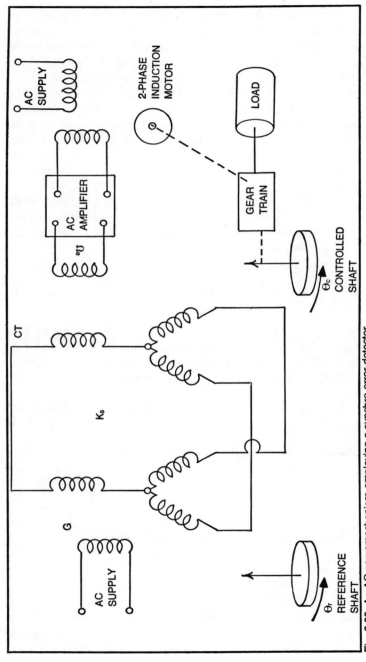

Fig. 6-33. An AC servomechanism employing a synchro error detector.

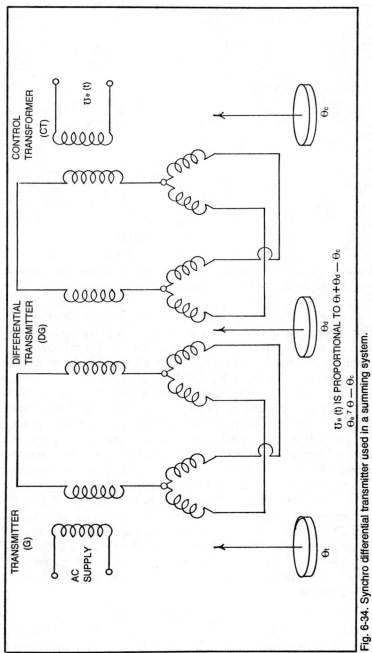

Fig. 6-34. Synchro differential transmitter used in a summing system.

193

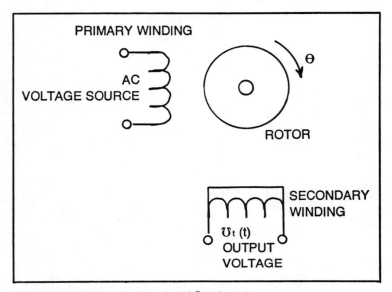

Fig. 6-35. Schematic diagram of an AC tachometer.

the stator of the synchro transmitter or the control transformer. The differential transmitter is often used to place a shaft in a position which is the sum of two shaft angles. For this purpose the unit is connected between a synchro transmitter and control transformer, as shown in Fig. 6-34.

The next error-sensing device to be discussed is called a *tachometer*. A tachometer is an electromechanical device that converts speed into a voltage. The AC tachometer illustrated in Fig. 6-35 is very similar to a two-phase induction motor. A sinusoidal voltage of rated value is applied to the primary winding of the tachometer, setting up a flux in the tachometer. The secondary winding is placed at a 90-degree angle mechanically with respect to the winding so that when the rotor shaft is stationary, the output voltage at the secondary winding is zero. When the rotor shaft is rotated, the output voltage is closely proportional to the rotor velocity. The polarity of the voltage is dependent on the direction of rotation. Thus, the characteristic equation of an AC tachometer can be written by the following time equation:

$$v_t(t) = K_t \frac{d\Theta}{dt}$$ **Equation 6-62**

In the equation 6-62, $v_t(t)$ is the output voltage, Θ is the shaft position, and K_t is the sensitivity of the tachometer in volts per radian or volts per rpm, which is usually specified by the manufacturer of the tachometer.

Equation 6-62 can be written in the frequency domain. Then transfer function of the AC tachometer can be written as follows:

$$\frac{V_t(s)}{\Theta(s)} = sK_t \qquad \textbf{Equation 6-63}$$

where $V_t(s)$ = the output voltage of the AC tachometer in volts, $\Theta(s)$ = the shaft position in radians, K_t = the AC tachometer sensitivity specified by the manufacturer in volts per radian or volts per rpm. Ranges for K_t are between 0.3V/1000 rpm to 10V/1000 rpm.

In servomechanisms, the applications of the AC tachometer are quite extensive. A system with *tachometer feedback* is known to have better stability in its response.

The DC tachometer of Fig. 6-36 is used in DC servo systems because the tachometer output is a DC voltage. The

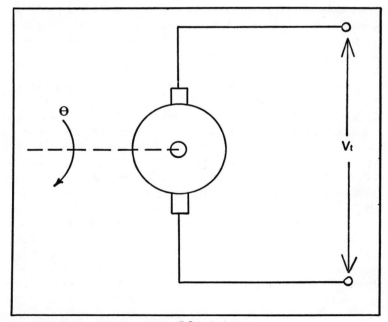

Fig. 6-36. Schematic diagram of a DC tachometer.

magnetic field of the device is obtained by means of a permanent magnet; therefore, no excitation voltage is required. The transfer of the DC tachometer is defined by the same equation as the AC tachometer; that is, Equation 6-63 is the transfer function for a DC tachometer as well as an AC tachometer.

A DC tachometer can also be used in an AC servo system if a modulator is used to convert its DC output signal into an AC voltage. Similarly, an AC tachometer can also be used in a DC servo system if a phase-sensitive demodulator is used to convert its AC output signal into a DC voltage.

TRANSDUCERS

A transducer is a device that receives a signal in one form (e.g., magnetic) and converts it to a signal in a different form (e.g., electric). The American Standards Association (ASA) defines a transducer as a device that "receives information in the form of a physical quantity and transmits this information in the form of another physical quantity." From the ASA definition, we can conclude that every component in a control system is a transducer. However, in a more limited usage, the term transducer refers to those devices used to measure some physical quantity, such as temperature, pressure, torque, displacement, force, velocity, etc. In this sense, *only components which provide the measuring means in a control system are considered transducers*.

Most transducers consist of two distinct parts: a primary element and a secondary element. The *primary element* is a device in which some physical quantity is dependent upon the value of the measured variable. The dependent physical quantity may be a displacement, a force, an electrical voltage, or an electrical resistance. The *secondary element* may be an amplifier, a transmitter, converter, discriminator, or detector. Its purpose is to convert the output of the primary element into a suitable electric or pneumatic signal.

In a control system, the function of the transducer is to provide a signal that is a measure of the controlled variable. For this reason, transducers are classified by the variable that is measured, rather than by the manner in which the output is developed. For example, in the previous section we illustrated error-sensing devices, which in fact illustrated devices that can be classified as transducers. We will proceed along the same lines throughout the remainder of this chapter.

Force Transducers

Most force transducers employ a means of producing a measurable balancing force. Two general methods are used to produce the balancing force: the null balance method and the displacement method. A *beam balance* is an example of a null balance force measuring means. The unknown mass is placed on one pan. Accurately calibrated masses of different sizes are placed on the other pan until the beam is balanced. The unknown mass is equal to the sum of the calibrated masses in the second pan.

A *spring scale* is an example of a displacement type of force measuring means. The unknown mass is placed on the scale platform which is supported by a calibrated spring. The spring is displaced until the additional spring force balances the force of gravity acting on the unknown mass. The displacement of the spring is used as the measure of the unknown force.

Three basic types of force transducers are described in this section. The *strain-gage load cell* and the volumetric load element are two displacement force transducers. For these two types, the unknown force is applied to an elastic member. The displacement of the elastic member is converted to an electric signal proportional to the unknown force. The third device is the *pneumatic force transmitter*. This is a null balance type. The unknown force is balanced by the force produced by air pressure acting on a diaphragm of known area. The air pressure is proportional to the unknown force and is used as the measured value signal.

Strain is the displacement per unit length of an elastic member. For example, if a bar of length L is stretched to length $L + \Delta L$, the strain, ϵ, is equal to $\Delta L/L$ by definition. A strain gage is a means of converting a small strain into a corresponding change in electrical resistance. It is based on the fact that the resistance of a fine wire varies as the wire is stretched (i.e., strained).

There are two general types of strain gages: bonded and unbonded. *Bonded strain gages* are used to measure strain at a specific location on the surface of an elastic member. The bonded strain gage is cemented directly onto the elastic member at the point where the strain is to be measured. The strain of the elastic member is transferred directly to the strain gage, where it is converted into a corresponding change in resistance. *Unbonded strain gages* are used to measure small displacements. A mechanical linkage causes the measured displacement to stretch a strain wire. The change in resistance of the strain wire is a measure of the displacement. The displacement is usually caused by a force acting on an elastic member. The unbonded strain gage measures the total displacement of an elastic member. The bonded strain gage measures the strain at a specific point on the surface of an elastic member.

The *gage factor* (G) of a strain gage is the ratio of the unit change in resistance to the strain. The gage factor can be expressed by the following equation:

$$G = \frac{\Delta R/R}{\Delta L/L} \qquad \text{Equation 6-64}$$

where G = the gage factor unitless quantity, ΔR = the change of resistance in ohms, R = the resistance of the strain gage, in Ohms, ΔL = the change in length in meters, and L = the length of the strain gage in meters.

The gage factor is usually between 2 and 4. The effective length (L) ranges from about 0.5 centimeters to about 4 centimeters. The resistance (R) ranges from about 50 ohms to 5000 ohms.

The stress on an elastic member is defined as the applied force divided by the unit area. If F is the applied force and A is the cross sectional area, the stress is equal to F/A. In elastic materials, the ratio of the stress over the strain is a constant called the *modulus of elasticity*, E, (see Table 6-1) defined by the following equation:

$$E = \frac{S}{\epsilon} \qquad \text{Equation 6-65}$$

where E = the modulus of elasticity in newtons per square meter, S = the stress in newtons per square meter, and ϵ = the strain in meters per meter.

Table 6-1. Modulus of Elasticity of Common Metals.

METAL	E in Newtons/square meter
Steel	$1.75 (10)^{11}$ to $2.2 (10)^{11}$
Aluminum	$5 25 (10)^{10}$ to $7.0 (10)^{10}$
Copper	$9.25 (10)^{10}$ to $1.3 (10)^{11}$

An example of a strain gage force transducer is shown in Fig. 6-37. The cantilever beam is the elastic member, and the unknown force is applied to the end of the cantilever beam, much like a diver on the end of a diving board. The strain produced by the unknown force is measured by a bonded strain gage cemented onto the top of the beam. The center of the strain gage is located a distance of L units from the end of the beam.

The cantilever beam assumes a curved shape that approximates a semicircle. The top surface is elongated, while the bottom surface is compressed. Halfway between these two surfaces is a neutral surface, in which there is no displacement.

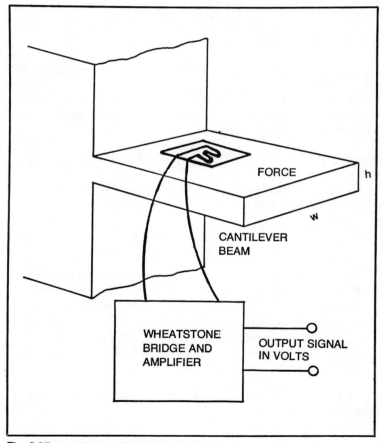

Fig. 6-37. A strain gage force transducer.

The stress at any point on the top surface of the cantilever beam is given by the following equation:

$$S = \frac{6\,FL}{wh^2} \qquad \textbf{Equation 6-66}$$

where S = the stress in newtons per meter, F = the applied force in newtons, L = the distance from the point the force is applied to any point on the top surface in meters, w = the width of the cantilever beam in meters, and h = the height of the cantilever beam, in meters.

We can combine Equations 6-64 through 6-66 to obtain an expression for the unit change in resistance produced by the unknown force. From Equation 6-64, it is known that strain $\epsilon = \Delta L/L$, so solve for the resistance ratio as follows:

$$\Delta R/R = G\epsilon \qquad \textbf{Equation 6-67}$$

Solving for the strain in Equation 6-65—that is, strain = S/E—and substituting this quantity into Equation 6-67, yields the following relationship:

$$\Delta R/R = GS/E \qquad \textbf{Equation 6-68}$$

At last, we substitute S in Equation 6-66 into Equation 6-68 to obtain the resistance ratio:

$$\frac{\Delta R}{R} = \frac{6GLF}{wh^2\,E} \qquad \textbf{Equation 6-69}$$

where ΔR = the change in resistance of the strain gage in ohms, R = the unstrained resistance of the strain gage in ohms, G = the gage factor of the strain gage (no units), L = the distance from the center of the strain gage to the end of the beam in meters, w = the width of the beam in meters, h = the height of the beam in meters, and E = the modulus of elasticity of the beam in newtons per square meter.

☐ **Example 6-5:** The strain gage force transducer in Fig. 6-37 has the following specifications.
    Material: Steel
    $E = 2\,(10)^{11}$ Newtons/square meter

Maximum allowable stress: 3.0 $(10)^8$ Newtons/square meter

w: 0.02 meters
h: 0.002 meters
L: 0.04 meters
G: 2
R: 100 Ohms

Determine the maximum force which can be measured and the change in resistance produced by the maximum force.

The maximum force is found through Equation 6-66 as follows:

$$F_{MAX} = \left| \frac{S_{MAX} \, w \, h^2}{6L} \right.$$

$$= \left| \frac{3 \, (10)^8 \, (0.02) \, (0.002)^2}{(6) \, (0.04)} \right.$$

$$= \left| 100 \text{ newtons} \right.$$

The change in resistance is obtained from Equation 6-69.

$$\Delta R = \frac{6RGLF}{wh^2E}$$

$$= \frac{(6) \, (100) \, (2) \, (0.04) \, (100)}{(0.02) \, (0.002)^2 \, (2) \, (10)^{11}}$$

$$= 0.6 \text{ ohms}$$

The *pneumatic force transducer* is shown in Fig. 6-38. The unknown force, F, is balanced by the force of the air pressure

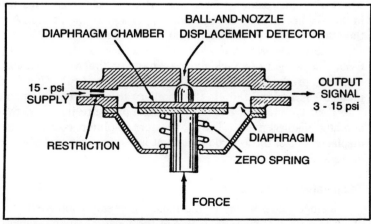

Fig. 6-38. A pneumatic force balance force transducer.

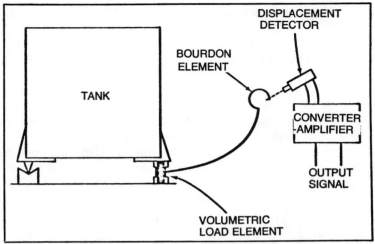

Fig. 6-39. A volumetric force transducer used to measure the weight of a tank.

against the effective area of the diaphragm. The ball and nozzle is arranged such that the balance of the two forces is automatic. Suppose force F increases. Then the force rod moves upward, reducing the opening between the ball and nozzle. The pressure in the diaphragm chamber increases and restores the balanced condition. The air pressure signal, P, in the diaphragm is determined by the following equation:

$$F = (P - 3)A \qquad \textbf{Equation 6-70}$$

where F = the unknown force in newtons, P = the air pressure in the diaphragm chamber in newtons per square meter, and A = the effective area of the diaphragm in square meters.

The *volumetric load element transducer* is shown in Fig. 6-39. The weight of a tank applies a force on the volumetric load element, producing a fluid pressure in the liquid fill. This fluid pressure is transmitted via the capillary to a Bourdon element. The Bourdon element converts the fluid pressure into a corresponding displacement. The converter amplifier converts the displacement of the Bourdon element into a proportional electric current signal.

Temperature Transducers

Temperature is one of the most commonly controlled variables in a control system. A few of the most common temperature transducers employed in control systems are the *thermocouple*, the *thermopile*, the *thermistor*, *thermoresistance*, and the *fluid expansion temperature transducer*.

The thermocouple shown in Fig. 6-40 consists of a junction of two dissimilar metals (iron and constantan). When the junction is subjected to a varying temperature, the free ends of the two metals produce a potential difference proportional to the temperature. The generated potential difference is rather small. For example, a copper-constantan thermocouple generates about 25 μV/°F. This low sensitivity implies the need for sophisticated amplification circuitry in order to obtain meaningful and usable voltages.

When selecting a thermocouple, consider not only the sensitivity but also the temperature range, as well as linearity over the temperature range, response time, etc. Typically, the linearity of thermocouples is poor (10 to 30 percent). It becomes necessary to tailor make amplifiers with a specific nonlinear response to compensate for thermocouple nonlinearity.

The time constant of thermocouples (time required to reach 63 percent of the final full voltage output) varies with the construction. Typical time constants range between 0.2 and 1.5 seconds. The thermocouple can be considered to be a temperature dependent voltage source with a very small output resistance close to an ideal voltage source.

Both the thermistor and thermoresistor function as temperature-dependent resistances. The thermistor consists of

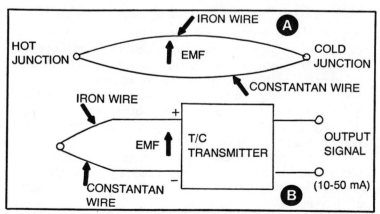

Fig. 6-40. A typical iron-constantan thermocouple at A, and a temperature control transmitter at B.

semiconductor material. The resistance of this material increases for decreasing temperatures; it has a negative temperature coefficient. The resistance is thus a measure of the temperature. Similarly, the thermoresistor consists of a wire with a positive temperature coefficient. Again the resistance is directly related to temperature.

To obtain an accurate reading of thermistor resistance versus temperature, it is simply necessary to measure resistance accurately as temperature changes. This can be done by the use of a *Wheatstone bridge*. The deviation from a null condition in the bridge is a measure of the sensitivity of the device. Therefore, in a precision temperature system where a thermoresistor is employed as the input transducer, the system designer usually finds his own data for the particular input transistor he employs.

The bimetallic thermostat is frequently used in control systems as an on-off temperature control, and it is illustrated in Fig. 6-41. The bimetallic device consists of two strips of different metals bonded together to form a leaf, or coil. The two metals must have different coefficients of thermal expansion so that a change in temperature will deform the original shape. A bimetallic thermometer is formed by attaching a scale and indicator to the bimetallic element such that the indicator displacement in proportional to the temperature. When the bimetallic device is open, this simulates an open switch in an electric circuit. When the temperature changes, the contacts will close, simulating a closed switch in an electric circuit.

Two additional types of temperature sensors are the *crystal-controlled oscillator* and *semiconductor junction*. The first is based on the fact that the resonant frequency of a crystal is temperature dependent. An increase in temperature increases the dimensions of the crystal and hence decreases its self-resonant frequency. An oscillator whose frequency is controlled by such a crystal will produce an output with a frequency inversely proportional to temperature. As temperature increases, frequency will decrease. With the oscillator designed to oscillate at a high frequency, a large frequency shift will result from a small temperature change, making this technique very sensitive and very suitable when small temperature variations are to be measured.

The forward voltage drop of a semiconductor junction is also temperature dependent. When we studied transistor theory,

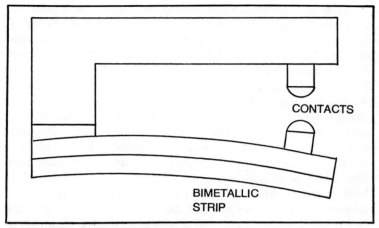

Fig. 6-41. A bimetallic thermostat.

we considered this temperature dependence a detrimental effect. However, the collector voltage changes with temperature can now be interpreted as a temperature indicator.

A gas filled fluid expansion temperature transducer is illustrated in Fig. 6-42A. The primary element consists of an inert gas sealed in a bulb which is connected by a capillary to a bellows pressure element. The bulb is immersed in the liquid to be measured until a thermal equilibrium is reached. The inert gas responds to the temperature change with a corresponding change in internal pressure. The primary element bellows convert the gas pressure into an upward force on the right-hand side of the force beam. The air pressure in the feedback bellows produces a balancing upward force on the left-hand side of the force beam.

The feedback bellows pressure is regulated by the combination of the restriction in the supply line and the relative position of the nozzle and force beam. Air leaks from the nozzle at a rate which depends on the clearance between the force beam and the end of the nozzle. The restriction in the supply line is sized such that the feedback bellows pressure is 3 psi when the nozzle clearance is a maximum. The bellows pressure will rise to 15 psi when the nozzle is completely closed by the force beam. The left end of the force beam and the nozzle form what is called a *flapper and nozzle displacement detector*. This is a very sensitive detector which produces a 3 to 15 psi signal, depending on a very small displacement of the flapper.

The arrangement of the force beam is such that it automatically assumes the position which results in a balance of forces. For example, assume a rise in the measured temperature occurred. The gas pressure increases in the primary element bellows, thereby increasing the force on the right-hand side of the force beam. The increased force tends to rotate the force beam about the flexure, which acts as a fulcrum. This moves the left end of the force beam closer to the nozzle, increasing the feedback bellows pressure. The increase in the feedback bellows pressure increases the balancing force on the left-hand side of the force beam. The force beam is balanced by a feedback bellows pressure which is always proportional to the primary element bellows pressure. The feedback bellows pressure is an accurate measure of the primary element gas pressure and is used as the output signal of the pressure transducer. A typical input output graph is included in Fig. 6-42B.

Flow Rate Transducers

In many control systems, the flow rate of liquids and gases is an important item to monitor. The measurement of the flow rate indicates how fluid or gas is used or distributed in a process. Flow rate is frequently used as a controlled variable to help maintain the efficiency and sometimes the economy of a given process. Today, flow rate can be analyzed by the help of computers, and adjustments made based on the data analysis of the computer.

The *average flow rate* is usually expressed in terms of the volume of liquid transferred in one second or one minute. The following equation defines flow rate:

$$Q = \frac{\Delta V}{\Delta t} \qquad \textbf{Equation 6-71}$$

where Q = the flow rate in cubic centimeters per second, ΔV = the change in volume in cubic centimeters, and Δt = the change in time in seconds.

The *instantaneous flow rate* is determined by taking the derivative of equation 6-71. It can be expressed mathematically as follows:

$$q = \frac{dV}{dt} \qquad \textbf{Equation 6-72}$$

where q = the instantaneous flow rate in cubic centimeters per second.

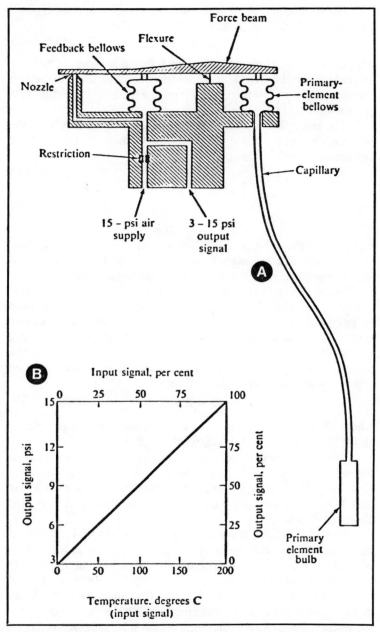

Fig. 6-42. A fluid-expansion temperature transducer at A, and its input-output graph at B.

The flow rate in a pipe is expressed in terms of fluid velocity x and cross sectional area A, as follows:

$$Q = Ax \qquad \text{Equation 6-73}$$

where Q = the flow rate of a pipe in cubic centimeters per second, A = the cross sectional area of the pipe in square centimeters, and x = the fluid velocity of the pipe in centimeters per second.

The *mass flow rate* of a fluid is obtained by multiplying the flow rate by the fluid density. It can be defined by the following equation:

$$M = \rho Q \qquad \text{Equation 6-74A}$$

$$m = \rho q \qquad \text{Equation 6-74B}$$

where M = the average mass flow rate in kilograms per second, m = the instananeous mass flow rate, ρ = the fluid density, Q = the average flow rate, and q = the instantaneous flow rate.

Differential pressure flow meters operate on the principle that a restriction placed in a flow line produces a pressure drop proportional to the flow rate squared. A differential pressure transmitter is used to measure the pressure drop (H) produced by the restriction. The flow rate (Q) is proportional to the square root of the measured pressure drop, which can be expressed by the following equation:

$$Q = K\sqrt{H} \qquad \text{Equation 6-75}$$

where Q = the flow rate, K = the volume of liquid per pulse, and H = the pressure drop across the orifice.

The restriction most often used for flow measurement is the orifice plate, which is a plate with a small hole, shown in Fig. 6-43. The orifice is installed in the flow line in such a way that all the flowing fluid must pass through the small hole also shown in Fig. 6-43.

Special passages transfer the fluid pressure on each side of the orifice to opposite sides of the diaphragm unit in a differential pressure transmitter. The diaphragm arrangement converts the pressure difference across the orifice into a force on one end of a force beam. A force transducer on the other end of the beam produces an exact counterbalancing force. A displacement detector senses any motion resulting from an inbalance of the forces on the force arm. The amplifier converts this displacement

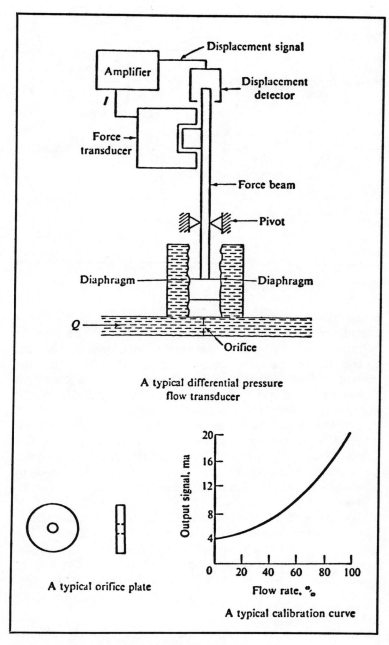

A typical differential pressure
flow transducer

A typical orifice plate

A typical calibration curve

Fig. 6-43. A differential pressure flow transducer.

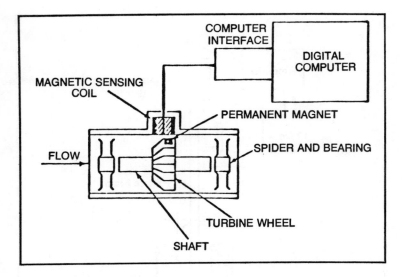

Fig. 6-44. Turbine flow meter.

signal into an adjustment of the current input to the force transducer that restores the balanced condition. The counter-balancing force produced by the force transducer is proportional to both the pressure drop and the input current (I). Thus, the current is directly proportional to the pressure drop across the orifice (H). This same electric current is used as the output signal of the differential pressure transducer.

In Fig. 6-43, the orifice is the primary element, and the differential pressure transmitter is the secondary element. The orifice converts the flow rate into a differential pressure signal, and the transmitter converts the differential pressure signal into a proportional electric current signal. A typical calibration curve is shown in Fig. 6-43.

A *turbine flow meter* is shown in Fig. 6-44. A small perma-nent magnet is imbedded in one of the turbine blades. The magnetic sensing coil generates a pulse each time the magnet passes by. The number of pulses is related to the volume of liquid passing through the meter by the following equation:

$$V = KN \qquad \textbf{Equation 6-76}$$

where V = the total volume of liquid, K = the volume of liquid per pulse, and N = the number of pulses.

The flow rate is equal to the total volume divided by the time interval. It can be expressed by the following equation:

$$Q = V/t = KN/t \qquad \textbf{Equation 6-77}$$

where Q = the flow rate, V = the total volume, t = the time interval, K = the volume of liquid per pulse, and N = the number of pulses.

The pulse output of the turbine flow meter is ideally suited for digital counting and control techniques. Digital blending control systems make use of turbine flow meters to provide accurate control of the blending of two or more liquids. Turbine flow meters are also used to provide flow rate measurements for input to a digital computer, as shown in Fig. 6-44.

The *magnetic flow meter* has no moving parts and offers no obstructions to the flowing liquid. It operates on the principle that a voltage is induced in a conductor moving in a magnetic field. A magnetic flow meter is illustrated in Fig. 6-45. The saddle-shaped coils placed around the flow tube produce a magnetic field at right angles to the direction flow. The flowing fluid is the conductor, and the flow of the fluid provides the

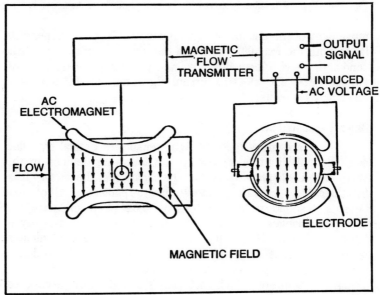

Fig. 6-45. Magnetic flow meter.

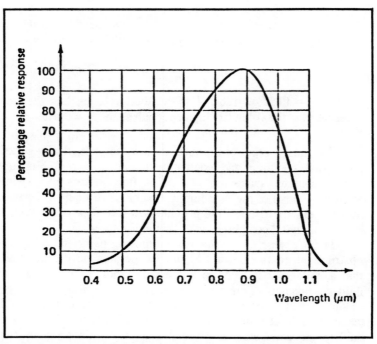

Fig. 6-46. Typical spectral response curve of a phototransistor.

movement of the conductor. The induced voltage is perpendicular to both the magnetic field and the direction of motion of the conductor. Two electrodes are used to detect the induced AC voltage into a DC electric current signal suitable for use by an electronic controller.

Optical Transducers

There are many types of light sensors, or optical transducers. We will limit ourselves to the following devices in this section: the *bulk photoconductor*, the *photodiode*, the *phototransistor*, the *photovoltaic cell (photocell)*, and the *phototube*. All of these devices respond to light intensity. The photoconductor, for example, changes resistance as a function of light intensity or light irradiation. This change corresponds to the light incident on the sensor. A large light intensity variation will produce a correspondingly large resistance change.

First, some of the characteristics of light will be covered. Light is an electromagnetic wave that has a distinct frequency and wavelength. The relationship between light frequency and wavelength are given by the following equation.

$$f = c/\lambda \qquad \text{Equation 6-78}$$

where f = the frequency of light in hertz; c = the velocity of light, which is $300\,(10)^6$ meters per second, and λ = The wavelength of light in meters.

When we discuss light, we must specify the frequency or the wavelength of light. The units we use are fractions of meters

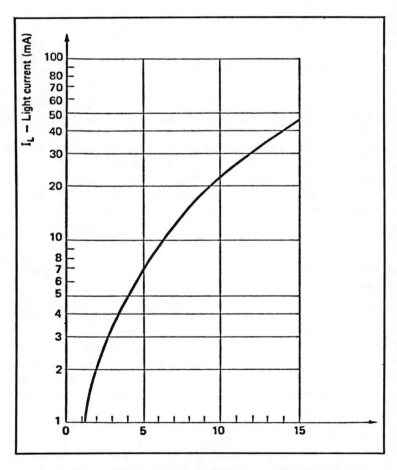

Fig. 6-47. Light current-versus-irradiance curve. Vce = 10V unfiltered tungsten filament bulb at 2870° K.

for the wavelength. One unit is called an *Angstrom* (A), which represents $(10)^{-10}$ meters.

The wavelength of visible light to the human eye ranges from 4000 to 7000 A. The eye, however, responds differently to the different wavelengths. For example, the eye is more sensitive to yellow light (5600 A) than it is to red light (6700 A). Similarly, the photosensor has a specific response curve. Figure 6-46 gives the spectral response curve of a silicon phototransistor. The particular transistor is most sensitive to light of 8000A.

Light is an electromagnetic wave that is a mode of propagating energy. When light impinges on a surface, we can talk about the energy transmitted to the surface per unit time or the power in watts impinging on the surface. The light sensitive device responds to this power.

Remember that the light power discussed here must be of a suitable wavelength. To obtain a more general criterion, we describe the response of the device to light exposure (irradiation) in terms of watts per unit area, typically mW per square centimeter. The behavior of a phototransistor can then be described in terms of a graph relating the light current versus the irradiance. Figure 6-47 shows the approximate light irradiation in the visible spectrum for various lighting conditions. This will permit you to appreciate the magnitudes involved in this discussion of light irradiation and photodevices.

A photoconductor consists of a material whose conductivity (or resistivity) changes with the intensity of light incident upon it. The photoconductor has a very high resistance (usually megohms) at very low light levels. Its resistance drops to a few kilohms when exposed to a light source. The more intense the light, the lower the resistance. When using a phototresistor, it is important to select the proper dark resistance as well as suitable sensitivity. Clearly, the spectral response of the sensor must match that of the light source. For example, if the light source to be used produces most of its light in the range of 0.5 to 0.6 μA, the light sensor should be most sensitive in this region in order to avoid a high degree of inefficiency. This holds true for all photosensitive devices.

The photoconductor usually has relatively a large sensitive area. A small light intensity change causes a large resistance change. It is quite common for a photoconductive element to exhibit a 1000-to-1 resistance change for a dark-to-light irradiance change of 0.5 μW per cm^2 to 5 mW per cm^2. The relation between irradiance and resistance is, however, not linear. It is close to exponential. This almost completely excludes the photoconductor from analog applications, where light intensity is the parameter to be measured.

The photoconductor has a number of other important characteristics. It has relatively slow response. A rise time of 1 mS is considered good, while 10 mS is quite common. At low light intensities (0.05 mW per cm^2), rise and fall times are in the order of 0.1 to 0.5 seconds. Units with response of 0.1 mS or less exhibit a substantial temperature instability. Because temperature changes cause substantial resistance changes for a constant light intensity, this type of photoconductor is unsuitable for analog applications.

A simple application of the photoresistor is shown in Fig. 6-48. The circuit is a light-controlled relay, which will energize by light and deenergize by the absence of light. The exact values of dark resistance and the resistance when subjected to light can be selected after you know the energizing voltage of the relay.

The photodiode is a reverse biased diode with a leakage current that is dependent on the intensity of the incident light. A set of typical characteristic curves is shown in Fig. 6-49. The currents involved are very small, measured in microamperes.

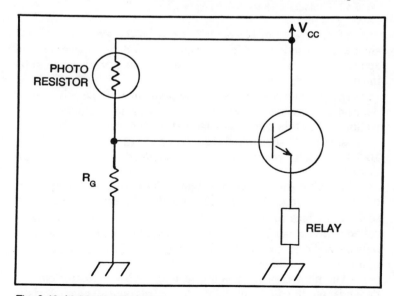

Fig. 6-48. Light-controlled relay.

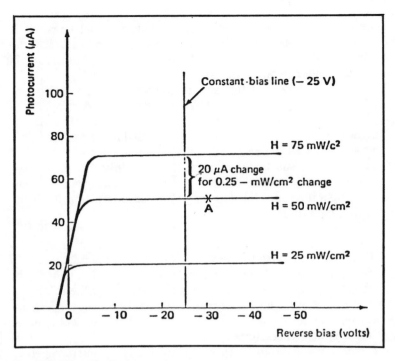

Fig. 6-49. Photodiode characteristic curves.

The effective area of the photodiode is on the average, $2 (10)^{-3}$ square centimeters, which makes it suited for space limited applications. The small size also results in a very small total incident light energy.

The frequency response of the photodiode is largely dependent on the intrinsic diode capacity. From semiconductor theory we know that this capacity is largest for low reverse bias and is very small for substantial reverse bias (2 pF at -10V). A 10-MHz cutoff frequency can be obtained employing photodiodes.

The photocurrent-versus-light relationship is quite linear over a wide range. In order to maintain this excellent linearity, to keep the bias voltage constant to operate the device in what is called a *current mode*. Note that even with bias voltage variations, only a small current variation will occur. This is a typical behavior of all standard diodes.

The phototransistor is a transistor whose collector current depends on the incident light. A set of typical characteristic curves is shown in Fig. 6-50. As seen, the curves are very much like those of a regular transistor; however, instead of I_B as a parameter, H, the light irradiance in mW per cm², is the running variable. In other words, the collector current depends upon H rather than I_B.

The phototransistor is mostly used in on-off applications because of its poor linearity and high sensitivity. Because the phototransistor is most suitable for digital applications, the rise and fall times are of major importance. Typically $t_r = 1\mu S$ and $t_f = 10\ \mu S$.

The photovoltaic cell produces a voltage proportional to light intensity. To obtain usable currents, the physical size of the photocell must be relatively large compared to the photoconductor or phototransistor. The latter two require an outside supply voltage, which determines the maximum available current. The photocell generates its own voltage and provides a limited current, usually of the order of 10 mA. This excludes applications where the photocell is designed to supply electrical power, such as in spacecraft applications or other solar cell applications. The photocell has

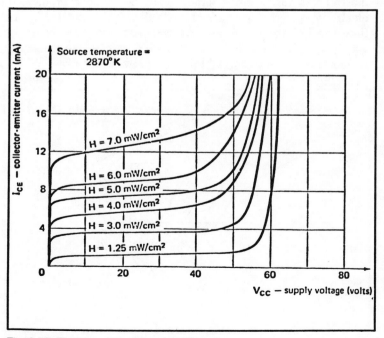

Fig. 6-50. Phototransistor characteristics.

a poor frequency response, low-output resistance, and good linearity of photocurrent-versus-radiation power.

The phototube relies on *photoemission*, the emission of electrons controlled by light. The cathode of the phototube is made of a photoemissive material. Thus, the cathode-to-plate current depends on the light incident on the cathode. The plate is usually returned to a high voltage. The currents are very small—on the order of mA. The sensitivity is high. An improvement on the phototube is the *photomultiplier*, which provides an extremely high sensitivity and can operate with extremely low light levels. It requires high supply voltages in excess of 1 kV. The photomultiplier is essentially a multiplate vacuum tube where the cathode is photoemissive. The first plate and the cathode comprise the phototube. The additional plates amplify the current produced by the phototube section. The current amplification, which ranges between 10^5 and 10^8, is attained by use of secondary emission. The secondary electrons emitted from the plate are accelerated and directed to the second dynode and so on. Since the number of secondary electrons is many times larger than the number of incident electrons, a very high current multiplication is obtained. The time it takes the electrons to travel from the cathode to the output dynode is on the order of 10 nS. Hence, the frequency response is on the order of a few megahertz.

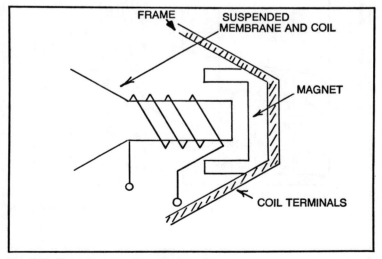

Fig. 6-51. Dynamic microphone construction.

Table 6-2. Microphone Characteristics.

TYPE	Impedance in kΩ	Frequency Response in Hertz	Sensitivity
Dynamic	50	100 to 10,000	−57 dB
Carbon	5	100 to 9,000	−100 dB
Crystal	1000	100 to 12,000	−55 dB
Capacitive	0.6	30 to 16,000	−70 dB

Acoustical Transducers

Devices that convert pressure waves into electrical energy are acoustical transducers. Typical examples of these devices are the *dynamic microphone,* the *crystal microphone, carbon microphone,* and the *capacitive microphone.*

The dynamic microphone is illustrated in Fig. 6-51. An AC voltage is generated when a coil is moved in a magnetic field. A membrane is mechanically linked to a moving coil surrounded by a permanent magnet, moving the coil across the magnetic field produced by the magnet. The voltage developed across the coil terminals is a direct result of the motion of the coil, which clearly depends on the force of the sound waves.

In a carbon microphone, the acoustic waves are used to compress powdered carbon. The resistance of the powdered carbon varies with the pressure put on it. A resistive variation is proportional to sound pressure. The variation in resistance can be converted to an AC current by applying a DC voltage across the microphone terminals. The change in resistance caused by the sound waves will produce variations in the current that are directly related to the sound.

The crystal microphone is based on the *piezoelectric effect.* A carbon crystal produces an electric voltage in one axis when a compression force is applied to another axis. The sound pressure wave exerts a compression force on one of the crystal directions. The voltage produced across the other axis is again directly related to the sound waves.

In a capacitive microphone, the pressure of the sound waves is used to move one of the plates of a parallel plate capacitor. Because the distance between the plates affects the capacitance, a variation of the capacitance is proportional to the pressure of the sound waves.

These microphones use distinctly different phenomena to obtain transduction. These phenomena may be used in other types of transducers to convert other energy forms to electrical energy. Each of the microphones have a number of distinct characteristics important to their use in a system. Table 6-2 shows three important characteristics for these microphones. The characteristics are referenced to 1 mW of input power to the microphones.

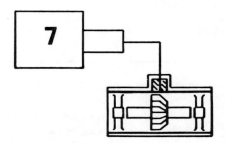

Practical Control Systems

In this chapter we will present a general mathematical approach to the analysis and design of common control systems. The models of control systems presented in this chapter will serve as a base from which a control system designer can begin. The designer must remember that most control system designs are specialized and the general design approach only carries the designer to a point where the designers imagination and intuition must be introduced into the control system design. In other words, the design of control systems contains both the scientific aspect and an art form, which involves the individuals designers imagination in facing the practical considerations of making the control system perform properly.

Figure 7-1 illustrates a typical open-loop control system block diagram. In the open-loop system, the output will follow the desired function as long as all the system variables are constant. Any change in load, amplifier gain, or any other system variable will cause a deviation from the desired value. The motor in the block diagram of Fig. 7-1 can control velocity, angular position, or torque. In order for the motor to conform to a desired function independently of changes in these variables, a closed-loop control system, such as the one illustrated in Fig. 7-2, must be constructed.

In the closed-loop system, the output variable is measured, fed back, and compared to the desired input function. Any difference between the two is a deviation from the desired

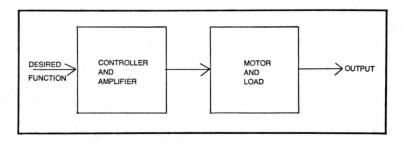

Fig. 7-1. Block diagram of an open-loop control system.

result; the deviation is amplified and used to correct the error. In this manner, the closed-loop system is essentially insensitive to variations in parameters and therefore performs correctly despite changes in load condition and other system parameters. However, now the response of the system depends on the closed-loop configuration, and as such it may be overdamped, underdamped, or even unstable. Special care must be given to the design of the closed loop in order to obtain the desired response.

Control systems are divided according to the variable being controlled. The most common control systems are:

□ Velocity Control System: A control in which motor velocity is to follow a given velocity profile.

□ Position Control System: A control system in which motor angular position is to be controlled.

□ Torque Control System: A control system in which motor torque is to be controlled.

□ Hybrid Control System: A control system in which the system switches from one control mode to another control mode. For example, you might want to control the velocity of the motor for a period of time and then switch to a position control mode. In this chapter, we will deal with the four basic control systems defined above, and spinoff control systems from the basic four that are commonly employed.

DC MOTOR CONTROL SYSTEMS

In this section, we will consider the three basic types of control systems employing a DC motor as the prime mover in the forward path of the control system, as shown in Fig. 7-3. The schematic diagram of an armature-controlled DC motor is

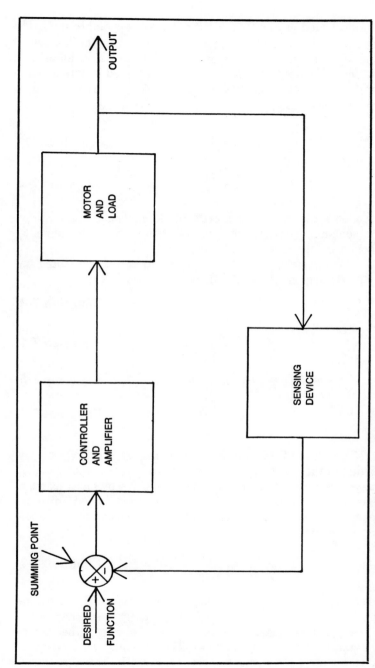

Fig. 7-2. Block diagram of a closed-loop control system.

223

also illustrated in Fig. 7-3. A DC voltage is applied to the field winding, and a variable voltage (v) is applied to the armature windings. The armature is represented by a resistor, an inductor, and an induced voltage source (v_i) connected in series. The armature current (i) is defined by the following frequency domain equation:

$$V = s(L)I + Ir + V_i \qquad \text{Equation 7-1}$$

$$I = (V - V_i)\ \frac{1}{sL + r} \qquad \text{Equation 7-2}$$

Equation 7-2 is represented on the block diagram of Fig. 7-3B by the summing junction, and the block between the summing junction and the current (I).

The remaining system equations are each represented on the block diagram of Fig. 7-3B, and are stated below:

$$T = K_T I \qquad \text{Equation 7-3}$$

$$W = \frac{T}{Js + b} \qquad \text{Equation 7-4}$$

$$V_i = K_E W \qquad \text{Equation 7-5}$$

$$\Theta = \frac{W}{s} \qquad \text{Equation 7-6}$$

The desired system transfer functions are obtained by solving in the system for the desired output over input ratio. If the velocity transfer function is desired, solve by block diagram algebra techniques for the ratio of armature speed/armature voltage, expressed mathematically as W/V, and whose equation is given below:

$$\frac{W}{V} = \frac{K_T/rb}{K_1 s^2 + K_2 s + K_3} \qquad \text{Equation 7-7}$$

where V = The armature voltage in volts, W = The armature speed in radians per second, K_1 = (J/b) (L/r) in seconds, K_2 = J/b + L/r in seconds, J = the moment of inertia of the load in kilograms-square meters, b = The damping resistance in newton-meters per radian per second, L = The armature inductance in henrys, r = The armature resistance in ohms, K_T = The torque constant in newton-meters per ampere, K_E = The emf constant in volts per radian per second, $K_3 = (rb + K_E K_T)/rb$, and s = The Laplace operator.

If the position transfer function is desired, solve for the ratio of armature position/armature voltage, expressed mathematically as Θ/V, and whose equation is given below:

$$\frac{\Theta}{V} = \frac{1}{s}\left(\frac{W}{V}\right) \qquad \textbf{Equation 7-8}$$

where Θ = The armature position in radians, and W/V = The transfer function for velocity given by Equation 7-7.

□ **Example 7-1:** Determine the velocity and position transfer functions of the block diagram of Fig. 7-3, given the following system parameters.

$$J = 6.2\ (10)^{-4}\ \text{kg-sq. meter}$$
$$b = 1\ (10)^{-4}\ \text{N-m/rad/S}$$
$$L = 0.02\ \text{H}$$
$$r = 1.2\ \text{ohms}$$
$$K_T = 0.06\ \text{N-m/A}$$
$$K_E = 0.06\ \text{V/rad/S}$$

First, find the various constants of the velocity transfer function given in Equation 7-7.

$$K_1 = (J/b)\ (L/r) = \frac{6.2\ (10)^{-4}\ (0.02)}{(10)^{-4}\ (1.2)}$$

$$K_1 = 0.103$$
$$K_2 = J/b + L/r = 6.2\ (10)^{-4}/(10)^{-4} + 0.02/1.2$$
$$K_2 = 6.22$$
$$K_3 = (rb + K_E K_T)/rb = 1 + K_E K_T/(rb)$$
$$K_3 = 1 + 0.06(0.06)/(1.2)\ (10)^{-4} = 1 + 30$$
$$K_3 = 31$$
$$K_T/(rb) = 0.06/1.2(10)^{-4} = 500$$

To find the velocity control transfer function, substitute the constants just above into Equation 7-7 to obtain the following result:

$$\frac{W}{V} = \frac{500}{0.103s^2 + 6.22s + 31}$$

225

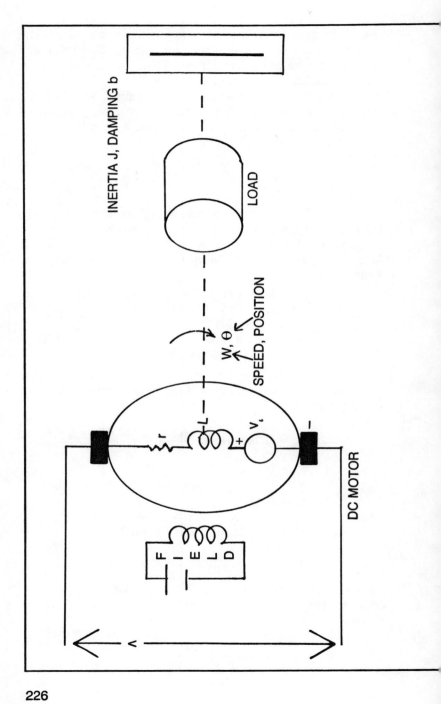

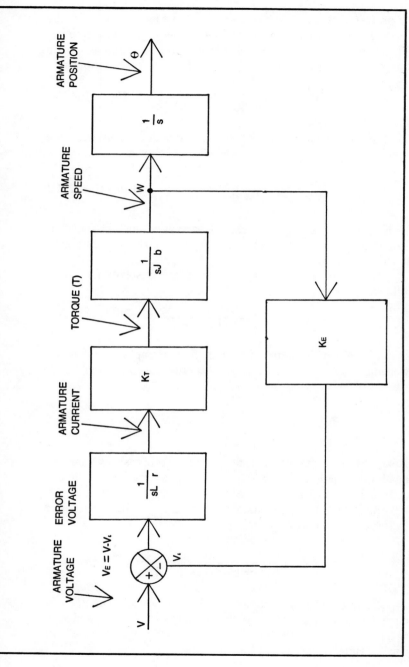

Fig. 7-3. DC motor control system schematic at A, and block diagram at B.

227

A preferred form of the above transfer function is found by making the coefficient of the highest power of s equal to unity. Thus, divide every term in the denominator by 0.103 and write the preferred form of the velocity transfer function as follows:

$$\frac{W}{V} = \frac{500}{s^2 + 60.39s + 301} \qquad \textbf{Equation 7-9}$$

To find the position control transfer function, substitute the result of Equation 7-9 into Equation 7-8 as follows:

$$\frac{\Theta}{V} = \frac{500}{s(s^2 + 60.39s + 301)} \qquad \textbf{Equation 7-10}$$

☐ **Example 7-2:** For the velocity transfer function found in Example 7-1, find the velocity response $\omega(t)$ to a unit step function input for V.

The result of Equation 7-9 is used and is solved for W, as follows:

$$W = \frac{500\,V}{s^2 + 60.39s + 301}$$

Since V is a unit step function in this example, the Laplace transform, 1/s, is substituted in the preceding equation for V. The result is shown below:

$$W = \frac{500}{s(s^2 + 60.39s + 301)}$$

Next, find the roots of the quantity in the parenthesis by using the quadratic equation, $(-b \pm \sqrt{b^2 - 4ac})/2a$, where b = 60.39, c = 301, and a = 1. The roots are found to be $r_1 = -54.91$ and $r_2 = -5.48$. Hence, the previous above equation can be written in a factored form as follows:

$$W = \frac{500}{s(s + 54.91)\,(s + 5.48)}$$

To find the time function for W, the velocity in the frequency domain, refer to Appendix A, transform No. 22, and let a = 54.91 and b = 5.48. Substitute these values in the transform pair for time. The result is the following:

228

$$\omega(t) = \frac{500}{(54.91)(5.48)}$$

$$\left[1 - \frac{(5.48) \, e^{-54.91t}}{(5.48 - 54.91)} + \frac{(54.91) \, e^{-5.48t}}{(5.48 - 54.91)}\right]$$

$$= 1.66(1 + 0.11 \, e^{-54.91t} - 1.11 \, e^{-5.48t})$$

$$= 1.66 + 0.1826 \, e^{-54.91t} - 1.8426 \, e^{-5.48t}$$

A graph of the velocity equation, $\omega(t)$ versus time, is shown in Fig. 7-4. Notice in Fig. 7-4 that after the two exponential functions decay to zero, the velocity of the motor settles to 1.66 rad/S.

AC MOTOR CONTROL SYSTEMS

Two-phase AC motors are often used in control systems which require a low-power, variable-speed drive. The primary advantage of the AC motor over the DC motor is its ability to use the AC output of synchros and other AC measuring means without demodulation of the error signal. An AC amplifier provides the gain for a proportional control mode. However, more elaborate control modes are difficult to implement with an AC signal. When additional control actions are required, the AC signal is usually demodulated, and the control action is inserted

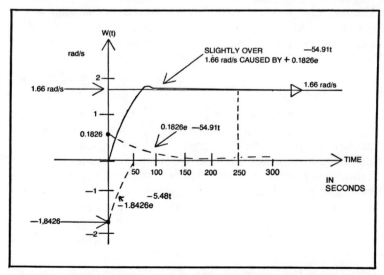

Fig. 7-4. The response of velocity $\omega(t)$ versus time for Example 7-2.

in the DC signal. The modified DC signal is then reconverted to an AC signal before the AC power amplifier is reconverted.

A very simple control system employing an AC motor is shown in Fig. 7-5. The input to this system is to be the position of the shaft on the input potentiometer, and the output is to be the position of the load shaft, which is sensed directly by the output potentiometer. The input potentiometer is excited with V_1 volts and has an active winding over Θ_1 radians of the shaft motion. Similarly, V_o and Θ_o describe the output potentiometer. The voltage summing circuit provides an error voltage, v_e, which is the difference between v_1 and v_o; v_e is therefore a function of the error or misalignment between the input shaft position, R, and the output shaft position, C. The exact relationship between v_e and the physical error or misalignment is easily derived from the following equations:

$$v_1 = R \left(\frac{V_1}{\Theta_1} \right)$$ Equation 7-11

$$v_o = C \left(\frac{V_o}{\Theta_o} \right)$$ Equation 7-12

$$v_e = v_1 - v_o$$ Equation 7-13

The amplifier boosts the level of v_e and supplies enough power to drive the AC servometer. Amplifier gain K_A is assumed to be a numeric (no time lag) and has the following equation:

$$K_A = \frac{v_m}{v_e}$$ Equation 7-14

A steady state speed torque curve for the motor is shown in Fig. 7-6. Although it is shown as a straight line, most servomotors have speed torque characteristics that are slightly curved, and the straight line drawn between the no-load speed point and the stall-torque point is an approximation made so the system can be treated as a linear one. From the speed torque curve, the *torque constant* for the motor is found by the following equation:

$$K_M = \frac{T_M}{V_M}$$ Equation 7-15

where K_M = the torque constant of the motor in dyne-cm per V,

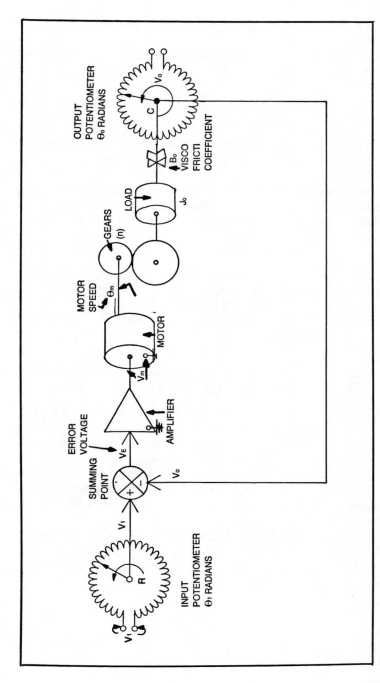

Fig. 7-5. AC motor control system.

231

T_M = the torque in dyne-cm, and V_M = the motor voltage in volts.

The equivalent viscous friction coeffieicnt is described by the following equation:

$$f_M \quad \frac{T_M}{\omega_M} \qquad \text{Equation 7-16}$$

where f_M = the viscous friction coefficient in dyne-cm per rad per S, T_M = the torque in dyne-cm, and ω_M = the motor speed in radians per second.

If the moment of inertia of the motor is J_M and the gear ratio between the motor shaft and output shaft is n, the shaft position of the motor can be described by the following equation:

$$\Theta_M = nC \qquad \text{Equation 7-17}$$

where Θ_M = the shaft position of the motor in radians, n = the gear ratio between the motor shaft and the output shaft (no units), and C = the output shaft position in radians.

The transfer function relating the output (C) and motor voltage (V_M) can be written as follows:

$$\frac{C}{V_M} = \frac{nK_M}{(J_o + n^2J_M)s^2 + (B_o + n^2f_M)s} \qquad \text{Equation 7-18}$$

where J_o = the moment of inertia of the output shaft in grams-square-cm, J_M = the moment of inertia of the motor shaft in grams-square-cm, and B_o = the viscous friction coefficient for the output shaft in dyne-cm per rad per S.

The other quantities in Equation 7-16 are described in Equations 7-17 and 7-16. The moments of inertia of the gears are not included in Equation 7-18, and all friction other than the viscous variety has also been neglected.

Equation 7-18 can be simplified to the following equation:

$$\frac{C}{V_M} = \frac{K}{s(s + a)} \qquad \text{Equation 7-19}$$

where:

$$K = \frac{nK_M}{J_o + J_M} \qquad \text{Equation 7-20}$$

$$a = \frac{B_o + n^2f_M}{J_o + n^2J_M}$$

Equation 7-21

The system which has been shown physically by Fig. 7-5 may be described analytically by a block diagram having a transfer function in each of the blocks. Such a diagram is shown in Fig. 7-7, where each block represents a particular physical element.

The block diagram of Fig. 7-7 can be rearranged into a simpler form, called the *unity feedback form*, without a change in its analytical meaning. This is done by employing the techniques of block diagram algebra presented in Chapter 2. The unity feedback block diagram is shown in Fig. 7-8. While preserving the analytical relationship between C(s) and R(s), the identity of each block has been destroyed. If the following definitions are made, the system transfer function can be defined in a preferred form for the unity feedback block diagram.

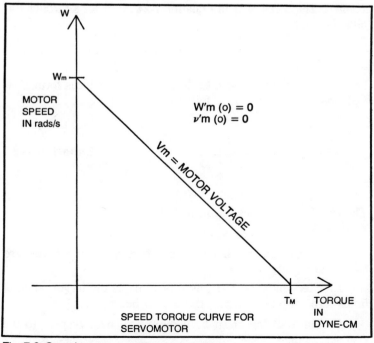

Fig. 7-6. Speed torque curve for a servomotor.

233

$$K_R = K_A K \left(\frac{V_o}{\Theta_o} \right)$$

Equation 7-22

$$K_P = \frac{V_1 \Theta_o}{\Theta_1 V_o}$$

Equation 7-23

$$W = \frac{C}{R} = \frac{K_P K_R}{s(s + a) + K_R}$$

Equation 7-24

Assume that all the physical parameters except the amplifier gain are fixed and that you wish to determine the way in which the dynamic response of the system to a step input changes K_A is varied. Amplifier K_A is not contained in K_P or in a, but is directly proportional to K_R, as shown in Equation 7-22. The locations of the poles of W will therefore change as K_A is varied. The root locus method is the most direct way of illustrating the manner in which the poles of W move around in the s-plane as K_A is changed. The poles of W in Equation 7-24 occur for s through the following equation:

$$s(s + a) + K_R = 0$$

This equation is rearranged to the standard root locus form, $1 + KGH = 0$ or $KGH = -1$, as follows:

$$-1 = \frac{K_R}{s(s + a)}$$

Equation 7-25

The root locus analysis begins as follows:

☐ The finite poles occur at $s = 0$ and $s = -a$.

☐ There are no finite zeros; however, two zeros are located at infinity.

☐ There must be as many root locus as the larger value of finite zeros or poles. Hence, there must be two root loci.

☐ The root loci must be symmetrical with respect to the real axis in the s-plane.

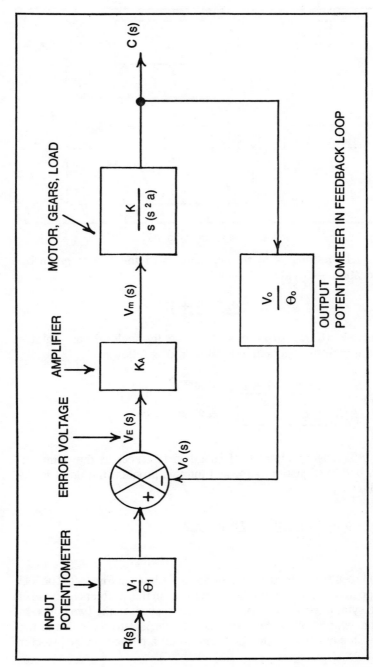

Fig. 7-7. Block diagram of the control system of Fig. 7-5.

235

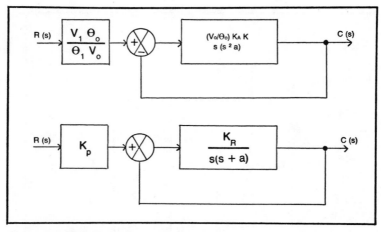

Fig. 7-8. Unity feedback block diagram of Fig. 7-7.

☐ The angles of the asymptotes of the loci is given by Equation 5-13 and shown below:

$$\beta = \frac{180° \, (2k + 1)}{n - m}$$

where $k = 0, 1, \dots (n - m - 1)$. Hence, the angles the asymptotes make with the real axis are calculated as follows:

$$\beta_1 = \frac{180°(2 \times 0 + 1)}{2} = 90°$$
$$\beta_2 = \frac{180° \, (2 \times 1 + 1)}{2} = 270° = -90°$$

☐ The asymptotes of root loci intersect at the point described by Equation 5-14, and for this problem the point is the following:

$$\sigma = \frac{(0 - a) - (0)}{2} = \frac{-a}{2}$$

The results obtained in these steps are shown in Fig. 7-9. The loci happen to coincide with the asymptotes in this problem because of the symmetry of the asymptotes with respect to the two poles. In other words, all of the root loci with respect to the poles of the transfer function in this problem lie on a perpendicular bisector with respect to the real axis. The perpendicular bisector happens to be the asymptote of the root locus method.

It is now possible to trace the path made by the poles of W as K_R is varied from zero to infinity. The magnitude requirement is described by the following equation:

$$1 = \left| \frac{K_R}{s(s+a)} \right|$$

Equation 7-26

which may also be written as:

$$K_R = |s| \, |s + a|$$

Equation 7-27

Equation 7-27 shows that for $K_R = 0$, $s = 0$ or $s = -a$ satisfy this equation and are therefore the poles of W for Equation 7-24.

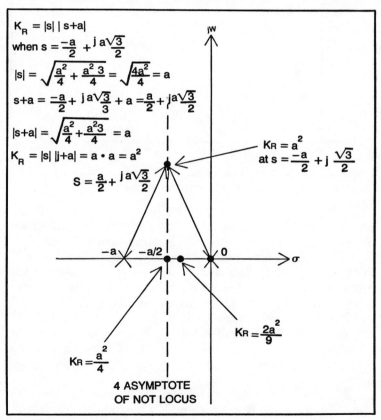

Fig. 7-9. Root locus diagram.

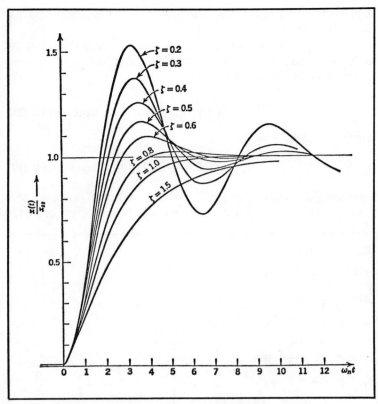

Fig. 7-10. Normalized second-order response curves.

If K_R is increased from zero to a small value, say $2a^2/9$, the two poles of W will lie at $s = -a/3$ and $s = -2a/3$, because these are the two values for s which satisfy Equation 7-27 when $K_R = 2a^2/9$. These two points are marked with dots in Fig. 7-9. For $K_R = a^2/4$, both poles of W lie at $s = -a/2$, as shown on the root locus plot of Fig. 7-9. As K_R is increased beyond $a^2/4$, the poles of W must move off the axis along the loci to points at which the magnitude requirement is satisfied. For $K_R = a^2$, for example, the poles of W lie at $s = -a/2 \pm ja\sqrt{3}/2$, because at these points, $|s| = a$ and $|s + a| = a$ satisfy Equation 7-27. As K_R in increased beyond a^2, the system poles move outward along the loci.

Because W has only two poles, it is apparent from the transfer function for W in Equation 7-24 that this control system is a second-order system. It is possible to write W in terms of a damping ratio and an undamped natural frequency. You can transform W into time by employing Transform No. 15 in Appendix A by corresponding the system parameters to the transform pair parameters through the following equations:

$$\omega_n^2 = K_R \; ; \; \text{and} \; \zeta = \frac{a}{2\sqrt{K_R}}$$

It is interesting to note that for critical damping ζ (zeta) = 1, K_R in this problem must be $a^2/4$. This checks with the root locus plot which shows a pair of poles at $s = -a/2$ for $K_R = a^2/4$.

Because the dynamic response of the second-order system is used extensively in control system engineering, it is essential to familiarize yourself with the previous response and pole locations of the second-order system described by Transform No. 15 in Appendix A. We will continue with our explanation of the AC motor control system model shown in Fig. 7-5 by assuming physical parameters for the control system and referring to the time response of the second-order system and its pole locations shown in Figs. 7-10 and 7-11 respectively.

☐ Motor Parameters:

T_M = 10 in-oz; W_M = 800 rpm; V_M = 100 V.

☐ Other System Parameters: J_m = 4 gram-sq.-cm; n = 10; J_o = 1 lb-sq.-cm; K_A = adjustable;

B_o = 0.5 in-oz per rad per S; V_o = 6V; Θ_o = 360°; V_1 = 6 V; Θ_1 = 180°.

Find amplifier gain K_A which will give the system an output response to a step input with approximately 5 percent overshoot. From the second-order time response curve in Fig. 7-10, a damping ratio (ζ) of about 0.7 will yield a 5 percent overshoot. From the pole zero location shown in Fig. 7-11, a damping ratio of 0.707 requires that the poles lie on the loci and on lines drawn through the origin, making angles of ±45° with the negative real axis. In other words, looking at Fig. 7-11, the cos $\Theta = \zeta = 0.707$, which means that the angle $\Theta = 45°$.

The system poles in this problem are located on the diagram shown in Fig. 7-12. The calculation of the loop system gain, K_R, can can now be made as follows:

$$K_R = |s| \, |s + a|$$

For value of $s = (-a/2 + ja/2)$, as shown on Fig. 7-12, the following calculations for the magnitudes of the factors s and s + a result:

$$|s| = \sqrt{\left(\frac{-a}{2}\right)^2 + \left(\frac{a}{2}\right)^2} = \sqrt{\frac{2a^2}{4}} = \sqrt{\frac{a}{2}}$$

$$s + a = \frac{-a}{2} + \frac{ja}{2} + a = \frac{a}{2} + j\frac{a}{2}$$

$$|s + a| = \sqrt{\left(\frac{a}{2}\right)^2 + \left(\frac{a}{2}\right)^2} = \sqrt{\frac{a}{2}}$$

$$K_R = |s|\,|s + a| = \left(\sqrt{\frac{a}{2}}\right)\left(\sqrt{\frac{a}{2}}\right)$$

$$K_R = \frac{a^2}{2}$$

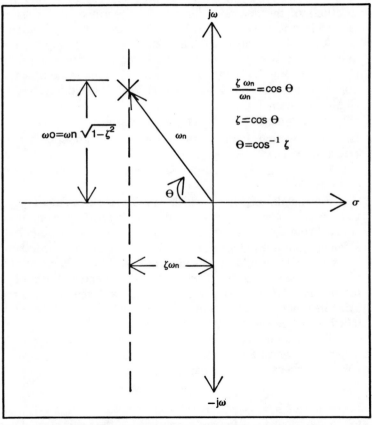

Fig. 7-11. Pole locations for transform No. 10, $1/s^2 + \zeta\omega_n s + \omega_n^2$.

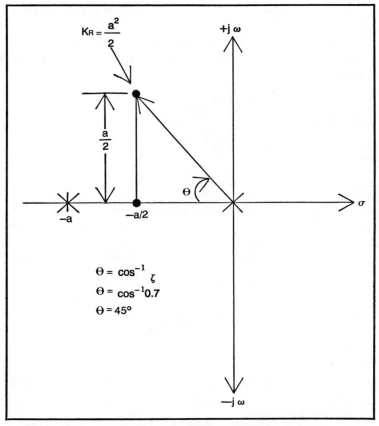

Fig. 7-12. Root locus plot when second-order system has $\zeta = 0.7$.

It is now necessary to find a numerical value for a from the physical parameters of the system. However, you must adopt a consistent set of units and express the physical parameters in that set. With grams as the unit of mass, dynes as the unit of force, centimeters for length, seconds for time, and radians for angular measurement, the system parameters given before can be described as follows:

☐ Motor Parameters:

$T_M = 705,000$ dyne-cm; $\omega_M = 837$ rad/S; $V_M = 100V$.

☐ Other System Parameters:

$J_M = 4$ gram-sq.-cm; $n = 10$; $J_o = 454$ gram-sq.-cm; $K_A =$ adjustable;

B_o = 35,400 dyne-cm per rad per S; V_o = 6 V; Θ_o = 2πrad;
V_1 = 6V; Θ_1 = π rad.

From these values and employing Equations 7-15, 7-16, 7-20, 7-21, and 7-22, you can solve for f_M, K_M, K, a, and K_A as follows:

$$K_M = \frac{T_M}{V_M} = \frac{705,000}{100}$$

$$K_M = 7050 \text{ dyne-cm/V}$$

$$f_M = \frac{T_M}{\omega_M} = \frac{705,000}{837}$$

$$f_M = 842 \text{ dyne-cm/rad/S}$$

$$K = \frac{nK_M}{J_o + n^2 J_M} = \frac{10(7050)}{454 + (10)2 \ (4)}$$

$$K = 82.5 \ / \ V\text{-s}^2$$

$$a = \frac{B_o + n^2 f_M}{J_o + n^2 J_m} = \frac{35,400 + (10)^2 \ (842)}{454 + (10)^2 \ (4)}$$

$$a = 140 \ /S$$

It is now necessary to calculate the loop gain, K_R, for this system from the information obtained by the root locus plot when $\zeta = 0.7$.

$$K_R = a^2/2 = (140)^2/2$$

$$K_R = 9800$$

$$K_A = \frac{K_R \Theta_o}{K V_o} = \frac{9800(2\pi)}{82.5(6)}$$

$$K_A = 124 \text{ V/V}$$

$$K_P = \frac{V_1 \Theta_o}{\Theta_1 V_o} = \frac{6(2\pi)}{\pi(6)}$$

$$K_P = 2$$

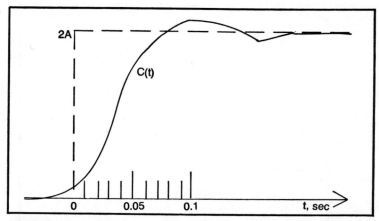

Fig. 7-13. Output response of second-order servomechanism to step input of A radians.

If a step input of A radians is applied to the input potentiometer shaft, the resulting output shaft motion will have the same form as the curve in Fig. 7-10 for $\zeta = 0.7$. Because the output shaft must move twice as far as the input shaft in order to balance the voltage of the input potentiometer, the steady state value for the output will be K_pA, or 2A. The response of the output shaft for the step input A is shown in Fig. 7-13. The time scale is determined for $\omega_n^2 = K_R$ through Equation 7-24 and Transform No. 15 in Appendix A. The actual value for $\omega_n = \sqrt{9800} \cong 99$.

As a second example, assume that in the control system previously described the amplifier is replaced by an amplifier which has a time lag of 1/b seconds and a steady state gain of K_1/b. A block diagram for the modified control system is shown in Fig. 7-14, where the transfer function of the new amplifier is given. The block diagram can be simplified to a unity feedback diagram by a manipulation of the block shown in Fig. 7-15.

If the time constant of the amplifier is 5 ms, then b = 1/5 mS = 200; and the system transfer function may be written as follows:

$$W(s) = \frac{2K_R}{s(s + 140)(s + 200) + K_R}$$

$$W(s) = \frac{2K_R}{s^3 + 340s^2 + 28000s + K_R} \qquad \textbf{Equation 7-28}$$

243

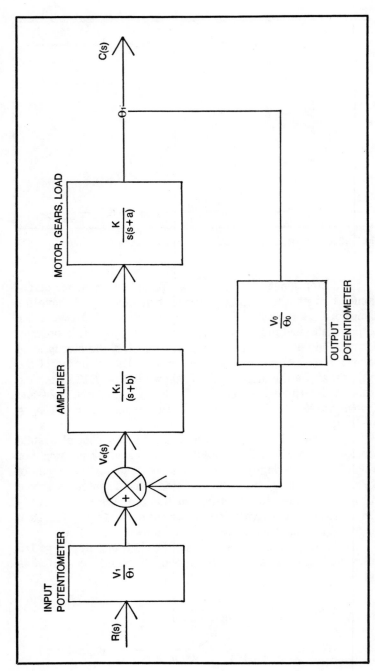

Fig. 7-14. Block diagram of a third-order control system.

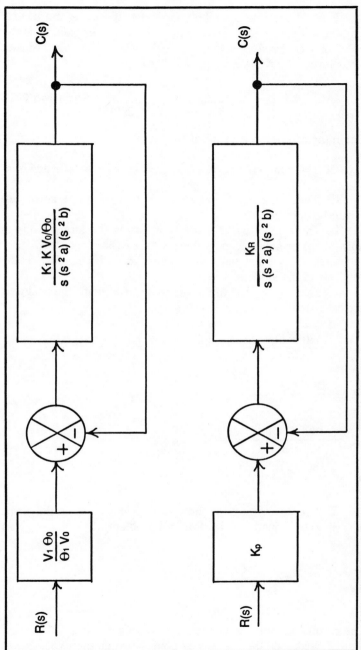

Fig. 7-15. Simplified block diagram for a third-order control system.

245

Now, determine how the transient response of the control system will vary with changes in K_1, the amplifier gain. Find the poles of W(s) by rearranging Equation 7-28 to the standard root locus form, $1 + KGH$ or $KGH = -1$, as follows:

$$-1 = \frac{K_R}{s(s + 140)(s + 200)} \qquad \text{Equation 7-29}$$

Begin the root locus analysis as follows:

☐ The finite poles occur at $s = 0$, $s = -140$, and $s = -200$.

☐ There are no finite zeros; however, three zeros are located at infinity.

☐ There must be as many root loci as the larger value of finite zeros or poles. Hence, there must be three root loci.

☐ The root loci must be symmetrical with respect to the real axis in the s-plane.

☐ The angles that the asymptotes make with the real axis are calculated as follows:

$$\beta = \frac{180° (2k + 1)}{n - m} \quad , \text{where } k = 0, 1, \dots (n - m - 1)$$

$$\beta_1 = \frac{180° (2 \times 0 + 1)}{3} = 60° \qquad \text{Equation 5-13}$$

$$\beta_2 = \frac{180° (2 \times 1 + 1)}{3} = 180°$$

$$\beta_3 = \frac{180° (2 \times 2 + 1)}{3} = 300° = -60°$$

☐ The asymptotes of the root loci intersect at the point described by Equation 5-14. For this problem, the point is given as follows:

$$\sigma = \frac{(0 - 140 - 200) - (0)}{3} = -113.33$$

The results obtained in these steps are shown in Fig. 7-16.

☐ Next, find the breakaway point between the two adjacent

poles $s = 0$, and $s = -140$ by using Equation 5-16, as shown in the following equation:

$$\frac{-1}{200 - c} \quad - \quad \frac{1}{140 - c} \quad = \quad \frac{-1}{c - 0}$$

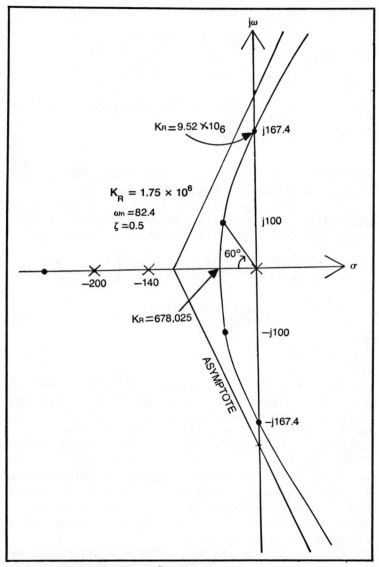

Fig. 7-16. Complete root locus diagram.

Solving the equation for breakaway point c, you will find that the breakaway point occurs at c $\cong$ -54.

☐ The value of K_R can be found at the breakaway point by substituting the value of s = c = -54 into Equation 7-28 and solving for K_R as follows:

$$-1 = \frac{K_R}{c(c + 140)(c + 200)}$$

$$K_R = (-c)(c + 140)(c + 200)$$

$$K_R = (-1)(-54)(-54 + 140)(-54 + 200)$$

$$K_R \cong 678,000$$

You find the value of K_R at the breakaway point, since K_R has its largest value on the real axis at the breakaway point.

☐ To find the point where the root loci cross the jω axis, apply the R-H test outlined in Chapter 5 in order to find the critical value for K_R. The R-H test yields the following array:

s^3	1	2800
s^2	340	K_R
s^1	$\dfrac{1(K_R) - 340(2800)}{340}$	0
s^0	2800	0

For the transfer function of Equation 7-28 to be stable, the coefficients of the first column in the array must all be positive numbers. In order to have a positive number for the s^1 term in the first column, K_R must be greater than the product, 340(2800), which is equal to 952,000. If K_R is exactly equal to 952,000, you have a conditionally stable point, which is the point where the root loci crosses the jω axis. If you substitute K_R = 952,000 into Equation 7-29 and solve for s, you will have the exact value for s that defines the crossing point of the root loci on the jω axis. The crossover point in Equation 7-29 when K_R = 952,000 is found to be s = j167.4.

☐ Once the real axis loci, asymptotes, breakaway, and crossover points are plotted, the entire root locus plot can be completed by finding through a process of trial and error estimation, several additional points which satisfy the angle requirement. These points are then connected in a smooth curve. The complete root locus diagram is shown in Fig. 7-16.

☐ Finally, determine the proper amplifier gain for this system. It is desirable to use as high again as possible because this minimizes the static error due to friction and other miscellaneous torque disturbances on the output shaft. On the other hand, as the gain is increased toward the point of instability, the transient response becomes oscillatory. It is therefore necessary to establish a specification for the transient response and to find the maximum value of gain which fulfills that specification. Usually as a rule of thumb, the transient response to a unit step must not exceed 16 percent of the final value of the output of the system. The step response of a third-order system having two complex poles is very nearly the same as the step response of a second-order system, provided the single pole is far from the imaginary axis. A good rule of thumb to employ in this case is to have the single pole at least twice the distance from the breakaway point on the real axis. In this system, the single pole is the amplifier pole of −200, which is almost four times the distance from the breakaway point of −54.

You must now decide where the poles of W(s) ought to lie in order to satisfy the 16 percent overshoot requirement and also what K_R must be in order to put the poles in the desired location. This may be done in the following manner. It is seen from the root locus plot that for K_R less than 678,000, all the poles of W(s) will lie on the real axis. No finite zeros occur in W(s), so the output C(s) will have the following form:

$$C(s) = \frac{2AK_R}{s(s + r_1)\,(s + r_2)\,(s + r_3)} \qquad \textbf{Equation 7-30}$$

where A = the magnitude of the step function, r = the real number root(s) of Equation 7-30, and K_R = the loop gain of the system.

The inverse transform of Equation 7-30 has the form:

$$c(t) = K_0 + K_1 e^{-r_1 t} + K_2 e^{-r_2 t} + K_3 e^{-r_3 t} \qquad \textbf{Equation 7-31}$$

Constants K_0, K_1, K_2, and K_3 are the residues of C(s) at the four poles and may be evaluated by the methods shown in Chapter 3.

The output expressed by Equation 7-31 never overshoots; that is, $c(t) < = K_o$ for all values of t. Therefore, it will be possible to operate this system within the stated specification when K_R is greater than 678,000 or in other words, when two poles of W(s) are on the complex portion of the locus. With two complex poles, the output will overshoot by an amount depending upon the positions of the poles.

Consider the system when K_R is set so that the damping ratio of the complex poles is 0.5. From the root locus plot in Fig. 7-16, you can determine that when $K_R = 175,000$ the poles will occur at s $= -257$ and s $= -42 \pm j72$ for the damping ratio of 0.5. Remember that the step response of a third-order system is approximately equal to the step response of a second-order system, provided the single real pole is far from the imaginary axis, which is the case in this example. For the step input, the residue at the real pole will be considerably smaller than the residues at the other poles of C(s), and the time constant of the component in c(t) associated with the real pole will also be small. This means c(t) will be very close to the response of a second-order system whose damping ratio is 0.5. A look at Fig. 7-10 shows this to be a response having approximately 16 percent overshoot, which is the specification for this system. Hence, the proper value for loop gain for this system is close to $K_R = 175,000$.

A K_R of 175,000 corresponds to an amplifier gain of K_1/b as follows:

$$\frac{K_1}{b} = \frac{K_R \Theta_o}{K V_o b} = \frac{175,000 \, (2\pi)}{82.5(6) \, (200)}$$

$$= 11.11$$

In practice, the amplifier is ordinarily designed with an adjustable gain, because the value for gain which is determined by linear analysis is based on data that is usually not exact. The speed torque characteristics of small AC servomotors will vary with small fluctuations in the manufacturing process. In this system, the amplifier would probably be built so that its gain could be varied up to 50. After the control system is assembled, the exact value for amplifier gain would be determined by experiment. Because the linear analysis is only an approximate representation of the physical system, it need not always be as precise as the calculations have shown in the two preceding examples.

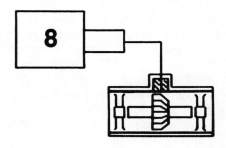

The Phase-Locked Loop

The phase-locked loop (PLL) is a closed-loop electronic servo system whose output locks onto and tracks an input reference signal. Phase lock is obtained by comparing the phase of the output signal with that of the reference, and any phase difference is converted into an error correction voltage. This error voltage changes the output signal phase to make it track the input.

The phase locked loop employs three basic electronic circuits. These are a *phase detector*, a *filter*, and a *voltage-controlled oscillator* (VCO). A block diagram of a basic PLL is shown in Fig. 8-1.

As the name implies, the PLL provides phase locking between two signals, that of the local VCO and an input signal. The phase-locked condition describes a fixed phase relation between the two signals. For example, the two voltages can be described by the following equations:

$$v_1 = A \sin(\omega t + \Theta_1) \qquad \textbf{Equation 8-1}$$

$$v_2 = B \sin(wt + \Theta_2) \qquad \textbf{Equation 8-2}$$

The phase difference between the two voltages in these equations is described by the equation;

$$\Theta_d = \Theta_1 - \Theta_2 \qquad \textbf{Equation 8-3}$$

To become familiar with the operation of the PLL, analyze the block diagram of Fig. 8-1 for constant Θ_d. Assume v_1 is the input or reference signal and the output voltage is v_2. The amplitude of the two voltages is of no importance. The VCO center frequency is adjusted to ω_c, the input frequency. Of course, all signal transformations are linear. The phase detector or comparator produces the phase difference Θ_d and converts it to a DC voltage. The conversion is linear; hence, we can talk about a constant voltage per degree of phase difference, $K_1 = V_d/\Theta_d$. The low-pass filter protects the system from any high-frequency disturbances, and the output of the low-pass filter will be V_d, which is fed into the amplifier with gain of K_2. Thus, the output of the amplifier is $V_c = K_1 K_2 \Theta_d$. The VCO is adjusted so that its center frequency ω_c is obtained while V_c is applied to it. The phase-locked loop is completed by V_2, which is the VCO output voltage, being fed back into the phase comparator. With sufficient loop gain ($K_1 K_2$), the VCO will be maintained at ω_c and phase difference Θ_d will be held constant. A change in Θ_1 will produce a change in Θ_d, which means that V_c will change, causing a new value for Θ_2. With Θ_2 being fed back to the phase detector, the new value of Θ_2 compensates for or corrects the change in Θ_1, again producing a constant phase error Θ_d.

When a programmable frequency divider is inserted into the feedback path, as shown in Fig. 8-2, the output becomes an integral multiple of the reference frequency. This technique is employed for multiple frequency generation in frequency synthesizers. The equation describing the output frequency is:

$$f_o = N f_r \qquad \textbf{Equation 8-4}$$

where f_o = the output frequency in hertz, f_r = the input or reference frequency in hertz, and N = the divide number (Mod No.) for the programmable frequency divider.

PLL DESIGN FUNDAMENTALS

The design of a PLL servo can be approached by employing the Laplace transform method of control systems. The parameters shown in Fig. 8-3 are defined for a PLL servo and will be used throughout this chapter. The Laplace transform permits the representation of a time response of a system, f(t), to be translated into the frequency response of a system, F(s). The response, F(s), is twofold in that it contains both the transient and steady state solutions. Thus, all operating conditions are

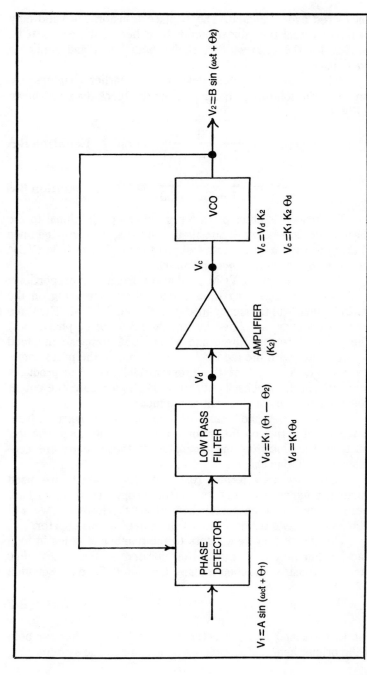

Fig. 8-1. Block diagram of the phase-locked loop.

253

considered and evaluated. The Laplace transform is valid only for positive real time linear parameters; hence, its use must be justified for the PLL which includes both linear and nonlinear functions.

Using a control system developed in earlier chapters, we can write the following equations from the block diagram shown in Fig. 8-3:

$$\Theta_e(s) = \frac{1}{1 + G(s)\,H(s)} \; \Theta_1(s) \qquad \textbf{Equation 8-5}$$

$$\Theta_2(s) = \frac{G(s)}{1 + G(s)\,H(s)} \; \Theta_1(s) \qquad \textbf{Equation 8-6}$$

The phase detector produces a voltage proportional to the phase difference between signals 0_1 and $0_2/N$. This voltage upon filtering is used as the control signal for the VCO/VCM (VCM means *voltage-controlled multivibrator*).

Because the VCO/VCM produces a frequency proportional to its input voltage, any time variant signal appearing on the control signal will frequency modulate the VCO/VCM. Thus, the output frequency is defined by Equation 8-4 during phase lock. The phase detector, filter, and VCO/VCM compose the feed forward path, with the feedback path containing the programmable divider. Removal of the programmable counter produces unity gain in the feedback path (N = 1). As a result, the output frequency is then equal to that of the input.

Various types and orders of loops can be constructed depending upon the configuration of the overall loop transfer function. Identification and examples of these loops are contained later in this chapter.

There are two terms, *type* and *order*, which are used somewhat indiscriminately in control theory. To date, in fact, there has not been an established standard. However, we will identify these two terms as they will be used in this chpater.

The *type* of a system refers to the number of poles in loop transfer function G(s) times H(s) located at the origin. For example, consider the loop transfer function defined by Equation 8-7.

$$G(s)\,H(s) = \frac{10}{s(s + 10)} \qquad \textbf{Equation 8-7}$$

The loop transfer function defined by Equation 8-7 has *one* pole at the origin; hence, Equation 8-7 defines a *type* one system.

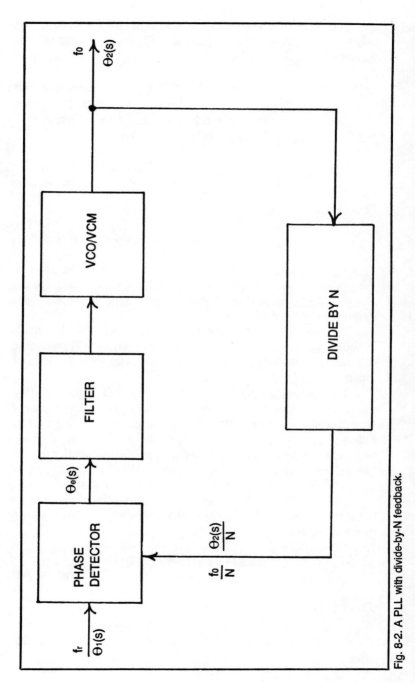

Fig. 8-2. A PLL with divide-by-N feedback.

255

The *order* of a system refers to the highest degree of the polynomial expression in Equation 8-8, which is termed the *characteristic equation* (C.E.):

$$1 + G(s)\,H(s) = 0 \qquad \textbf{Equation 8-8}$$

Using the loop transfer function defined by Equation 8-7, you can test for the order of the system as follows:

$$1 + G(s)\,H(s) = 1 + \frac{10}{s(s + 10)} = 0$$

Therefore:

$$\text{C.E.} = s(s + 10) + 10 = s^2 + 10s + 10 = 0$$

This equation is a second-*order* polynomial. Hence, the order of the system is two—that is, a second-order system. Thus, for the given G(s) H(s) defined by Equation 8-7, a type 1, second-order system is obtained.

Various inputs can be applied to a system. Typically, these include step position, velocity, and acceleration. The response of type 1, 2, and 3 systems will be examined with the various inputs.

To evaluate the system shown in Fig. 8-3, $\Theta_e(s)$ must be examined in order to determine if the steady state and transient characteristics are optimum and/or satisfactory. The *transient response* is a function of loop stability and will be covered shortly. The steady state evaluation can be simplified with the use of the final value theorem associated with Laplace transforms. This theorem permits finidng the steady state system error, $\Theta_e(s)$, resulting from input $\Theta_1(s)$ without transforming back to the time domain. The final value theorem simply stated is shown in Equation 8-9.

$$\underset{t \to \infty}{\text{Lim}}\ \Theta_e(t) = \underset{s \to 0}{\text{Lim}}\ s\,\Theta_e(s) \qquad \textbf{Equation 8-9}$$

where $\Theta_e(s)$ is defined by Equation 8-5.

For the typical inputs, you can characterize $0_1(s)$ as follows.

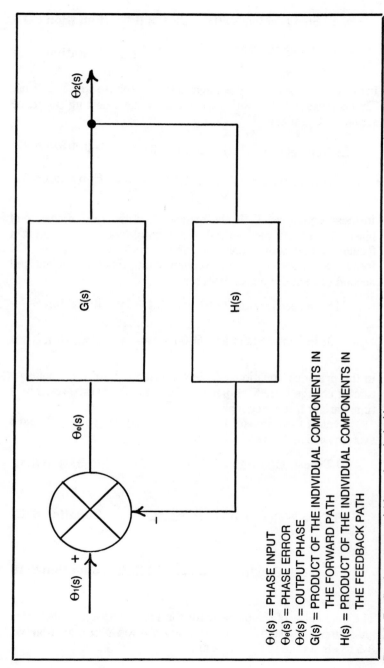

$\Theta_1(s)$ = PHASE INPUT
$\Theta_e(s)$ = PHASE ERROR
$\Theta_2(s)$ = OUTPUT PHASE
$G(s)$ = PRODUCT OF THE INDIVIDUAL COMPONENTS IN
      THE FORWARD PATH
$H(s)$ = PRODUCT OF THE INDIVIDUAL COMPONENTS IN
      THE FEEDBACK PATH

Fig. 8-3. Feedback system model for a phase-locked loop.

257

□ Step position: $\Theta_1(t) = C_p$; $t \geq 0$ **Equation 8-10**

Or in Laplace notation: $\Theta_1(s) = \dfrac{C_p}{s^1}$ **Equation 8-11**

In these equations, C_p is the magnitude of the phase step in radians. This corresponds to shifting the phase of the incoming reference signal by C_p radians.

□ Step velocity: $\Theta_1(t) = C_v t$; $t \geq 0$ **Equation 8-12**

Or in Laplace notation: $\Theta_1(s) = \dfrac{C_v}{s^2}$ **Equation 8-13**

In these equations, C_v is the magnitude of the rate of change of phase in radians per second. This corresponds to inputing a frequency that is different than the feedback portion of the VCO frequency. Thus, C_v is the frequency difference in radians per second seen at the phase detector.

□ Step acceleration: $\Theta_1(t) = C_a t^2$; $t \geq 0$ **Equation 8-14**

Or in Laplace notation: $\Theta_1(s) = \dfrac{2 C_a}{s^3}$ **Equation 8-15**

In the previous equations, C_a is the magnitude of the frequency rate of change in radians per second. This is characterized by a time variant frequency input.

Typical loop transfer function G(s) H(s) for types 1, 2, and 3 systems are as follows:

□ Type 1: $G(s) H(s) = \dfrac{K}{s(s + a)}$ **Equation 8-16**

□ Type 2: $G(s) H(s) = \dfrac{K(s + a)}{s^2}$ **Equation 8-17**

□ Type 3: $G(s) H(s) = \dfrac{K(s + a)(s + b)}{s^3}$ **Equation 8-18**

The final value of the phase error for a type 1 system with a step phase input is found by substituting Equation 8-11 into Equation 8-5 to obtain the following result:

$$\Theta_e(s) = \left[\cfrac{1}{1 + \cfrac{K}{s(s+a)}} \right] \frac{C_p}{s^1} \qquad \text{Equation 8-19}$$

$$\Theta_e(s) = \frac{(s+a)}{s^2 + as + K} C_p \qquad \text{Equation 8-20}$$

$$\Theta_e(\infty) = \lim_{s \to 0} s\, \Theta_e(s) = 0 \qquad \text{Equation 8-21}$$

Thus, the final value of the phase error is zero when a step position (phase) is applied. Similarly applying the three inputs into type 1, 2, and 3 systems and utilizing the final value theorem, Table 8-1 can be constructed showing the respective steady state phase errors.

A zero phase error identifies phase coherence between the two input signals at the phase detector. A constant phase error identifies a phase differential between the two input signals at the phase detector. The magnitude of this differential phase error is proportional to the loop gain and the magnitude of the input step. A continually increasing phase error identifies a time rate change of phase. This is an unlocked condition for the phase loop.

Using Table 8-1, the system type can be determined for specific inputs. For instance, if it is desired for a PLL to track a reference frequency (step velocity) with zero phase error, a minimim of type 2 is required.

PLL STABILITY FUNDAMENTALS

The root locus technique of determining the position of system poles and zeroes in the s-plane is often used to graphi-

Table 8-1. Steady State Phase Errors for System Types 1, 2 and 3.

Name	Type 1	Type 2	Type 3
Step Position	Zero	Zero	Zero
Step Velocity	Constant	Zero	Zero
Step Acceleration	Continually Increasing	Constant	Zero

cally visualize the system stability. The graph or plot illustrates how the closed-loop poles (roots of the characteristic equation) vary with loop gain. For stability, all poles must lie in the left half of the s-plane. The relationship of the system poles and zeros then determines the degree of stability. The root locus contour can be determined by using the following rules.

☐ Rule 1: The root locus begins at the poles of loop transfer function G(s) H(s) for K = 0 and ends at the zeros of G(s) H(s) for K = ∞, where K is the loop gain.

☐ Rule 2: The number of root loci branches is equal to the number of poles or number of zeros, whichever is greater. The number of zeroes at infinity is the difference between the number of finite poles and finite zeroes of G(s)H(s).

☐ Rule 3: The root locus contour is bounded by asymptotes whose angular position is given by Equation 8-22.

$$\phi = \frac{(2n + 1)\,\pi}{\#P - \#Z} \; ; \quad n = 0, 1, 2,\ldots \text{etc.} \quad \textbf{Equation 8-22}$$

where #P = the number of poles, and #Z = the number of zeroes.

☐ Rule 4: The intersection of the asympotes is positioned at the center of gravity (C.G.) and found by Equation 8-23.

$$\text{C.G.} = \frac{\sum P - \sum Z}{\#P - \#Z} \qquad \textbf{Equation 8-23}$$

where $\sum P$ denotes the algebraic sumation of poles, and $\sum Z$ denotes the algebraic summation of zeroes.

☐ Rule 5: On a given section of the real axis, root loci may be found in the section only if the #P+#Z to the right is odd.

☐ Rule 6: Breakaway points from the negative real axis is given by Equation 8-24.

$$\frac{dK}{ds} = 0 \qquad \textbf{Equation 8-24}$$

Again K is the loop gain variable factored from the characteristic equation.

An example of stability analysis follows employing the preceding six rules when applicable. Consider the root locus for a typical loop transfer function given in Equation 8-25.

$$G(s)\, H(s) = \frac{K}{s(s + 4)}$$ **Equation 8-25**

The root locus has two root loci branches which begin at the poles $s = 0$ and $s = -4$ and ends at the two zeroes located at infinity (Rule 1). The asymptotes can be found according to Rule 3. Because there are two poles and no zeroes, the equation becomes:

$$\text{For } n = 0,\ \phi_o = \frac{2(0) + 1}{2} = \frac{\pi}{2}$$

$$\text{For } n = 0,\ \phi_1 = \frac{2(1) + 1}{2} = \frac{3\pi}{2}$$

The position of the intersection according to Rule 4 is:

$$\text{C.G.} = \frac{\sum P - \sum Z}{+P - \#Z} = \frac{(-4 - 0) - (0)}{2 - 0)}$$

$$= -2$$

The breakaway point as defined by Rule 6 can be found by first writing the characteristic equation.

$$\text{C.E.} = 1 + G(s)H(s)$$

$$= 1 + \frac{K}{s(s + 4)} = s^2 + 4s + K = 0$$

Solving for K yields:

$$K = -s^2 - 4s$$

Taking the derivative with respect to s and setting it equal to zero then determines the breakaway point.

$$\frac{dK}{ds} = -2s - 4$$

$$0 = -2s - 4$$

$$s = -4$$

The value of s = −4 represents the point of departure. Using this information, the root locus can be plotted as shown in Fig. 8-4.

The second-order characteristic equation given by the following equation, $s^2 + 4s + K$, can be normalized to the standard form given by Equation 8-26.

$$s^2 + 2\zeta\omega_n s + \omega_n^2 \qquad \textbf{Equation 8-26}$$

The damping ratio $\zeta = \text{Cos } \phi(0° \leq \phi \leq 90°)$ and ω_n is the natural frequency as shown in Fig. 8-4.

The response of this type 1, second-order system to a step input is shown in Fig. 8-5. These curves represent the phase response to a step position (phase) input for various damping ratios. The output frequency response as a function of time to a

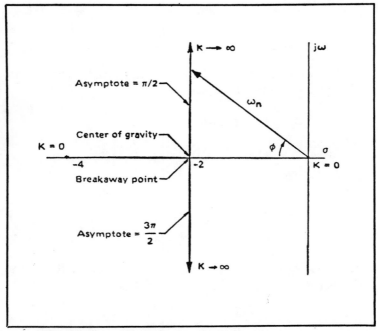

Fig. 8-4. Type 1 second-order root locus contour.

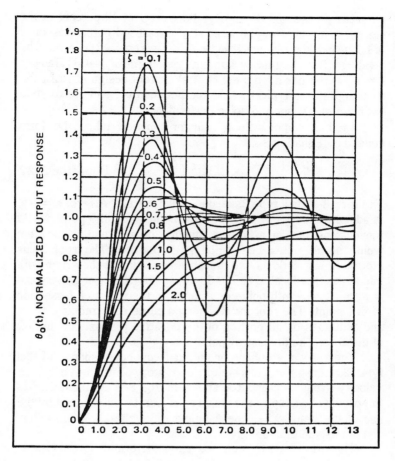

Fig. 8-5. Type 1 second-order step response.

step velocity (frequency) input is also characterized by the same set of curves shown in Fig. 8-6.

The overshoot and stability as a function of damping ratio ζ is illustrated by the various plots. Each response is plotted as a function of the normalized time, $\omega_n t$. For a given ζ and a lockup time t, the ω_n required to achieve the desired results can be determined as illustrated by the following example.

☐ **Example 8-1:** If the damping ratio is 0.5 and the allowable error is 10 percent or less, find ω_n for a lockup time of 1 mS.

From Fig. 8-5, find the curve expressed by a damping ratio of 0.5. The steady state value for this curve is a normalized output response of 1.0. An overshoot error of 10 percent of the steady state value would occur at a normalized output response of 1.1, which corresponds to $\omega_n t = 4.5$. The required ω_n is found by dividing 4.5 by the 1 mS lockup time yielding an $\omega_n = 4.5$ K rad per s = 4500 rad per s. In most control system design of PLL, the damping ratio is typically selected between 0.5 and 1.0 to yield optimum overshoot and noise performance.

Another common loop transfer function takes the form defined by Equation 8-27.

$$G(s)H(s) = \frac{(s + a) K}{s^2}$$

Equation 8-27

Equation 8-27 is a type 2, second-order system. A zero, $s = -a$, is added to provide stability. Without the zero, the poles would move along the $j\omega$ axis as a function of gain and the system would at all times be oscillatory in nature. The root locus plotted in Fig. 8-6 has two branches beginning at the origin with one asymptote located at 180 degrees. The center of gravity is s = a; however, with only one asymptote there is no intersection at this point. The root locus lies on a circle centered at $s = -a$ and continues on all portion of the negative real axis to the left of the zero. The breakaway point is $s = -2a$.

The respective phase or output frequency response of this type 2, second-order system to a step position (phase) or velocity (frequency) input is shown in Fig. 8-7. As illustrated in the previous example, the required ω_n can be determined by the use of the graph when the damping ratio, the allowable error, and the lockup time are given.

The -3 dB cutoff frequency can be determined for a type 1, second-order PLL system by Equation 8-28.

$$\omega_{-3dB} = \omega_n [1 - 2 \zeta^2 + \sqrt{2 - 4 \zeta^2 + 4 \zeta^4}]^{\frac{1}{2}}$$

Equation 8-28

The -3 dB cutoff frequency can be determined for a type 2, second-order PLL system by Equation 8-29.

$$\omega_{-3dB} = \omega_n [1 + 2 \zeta^2 + \sqrt{2 + 4 \zeta^2 + 4 \zeta^4}]^{\frac{1}{2}}$$

Equation 8-29

PHASE-LOCKED LOOP DESIGN EXAMPLE

The design of a PLL typically involves determining the type of loop required, selecting the proper bandwidth, and establishing the desired stability. A fundamental approach to these

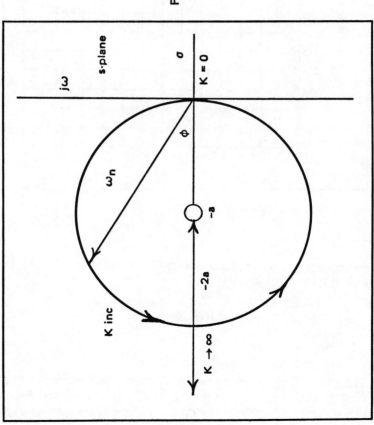

Fig. 8-6. Type 2 second-order root locus contour.

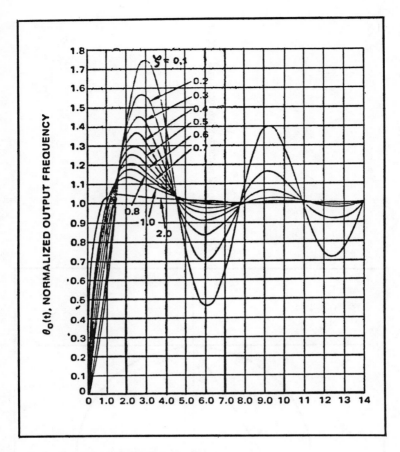

Fig. 8-7. Type 2 second-order step response.

constraints is already illustrated. It is desired for the system to
have the following specifications.

Output Frequency: 2.0 MHz to 3.0 MHz
Frequency Steps: 100 kHz
Lockup Time
Between Channels: 1 mS
Overshoot: Less than 20 percent

These specifications characterize a system function similar to a
variable time base generator or a frequency synthesizer. From
the given specifications, the circuit parameters shown in Fig.

266

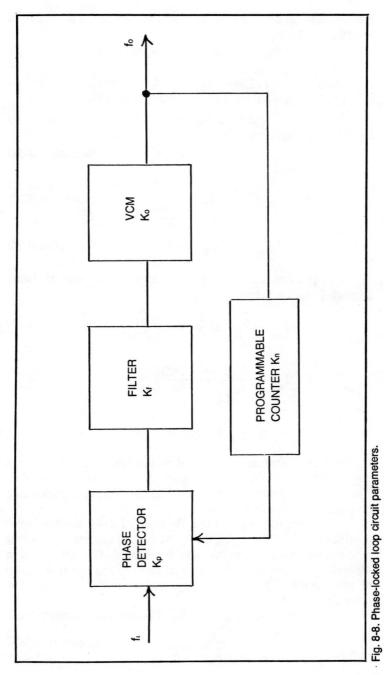

Fig. 8-8. Phase-locked loop circuit parameters.

8-8 can now be determined. The devices employed to construct the PLL are listed below.

Freqeuncy Phase Detector: MC4044/4344

Voltage-Controlled
Multivibrator (VCM): MC4024/4324

Programmable Counter: MC4016/4316

The forward and feedback transfer functions are given by the following equations.

$$G(s) = K_p K_f K_o \qquad \textbf{Equation 8-30}$$

$$H(s) = K_n \qquad \textbf{Equation 8-31}$$

$$K_N = 1/N \qquad \textbf{Equation 8-32}$$

The programmable counter divide ratio, K_n, can be found through Equation 8-4 as follows:

$$N_{min} = \frac{f_{o\ min}}{f_i} = \frac{f_{o\ min}}{f_{step}} = \frac{2\ \text{MHz}}{100\ \text{kHz}} = 20$$

$$N_{max} = \frac{f_{o\ max}}{f_{step}} = \frac{3\ \text{MHz}}{100\ \text{kHz}} = 30$$

$$K_n = \frac{1}{20}\ \text{to}\ \frac{1}{30}$$

A type 2 system is required to produce a phase coherent output relative to the input, as previously shown in Table 8-1. The root locus contour is shown in Fig. 8-6, and the system step response is shown in Fig. 8-7.

The operating range of the MC4024/4324 VCM must cover 2 MHz to 3 MHz. Selecting the VCM control capacitor according to the rules contained on the data sheet yields C = 100 pF. The desired operating range is then centered within the total range of the device. The input voltage versus output frequency is shown in Fig. 8-9.

The transfer function of the VCM is given by Equation 8-33.

$$K_o = \frac{K_v}{s} \qquad \textbf{Equation 8-33}$$

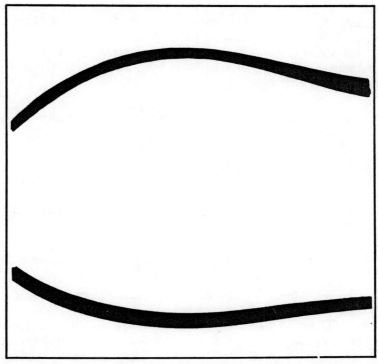

Fig. 8-9. Curves showing VCM control voltage transient.

The quantity, K_v, is the sensitivity in radians per second per volt. From the curve in Fig. 8-9, K_v is obtained by taking the reciprocal of the slope as shown in the following calculation:

$$K_v = \frac{4 \text{ MHz} - 1.5 \text{ MHz} (2\pi)}{5V - 3.6V} \quad \text{rad per S per V}$$

$$= 11.2 \, (10)^6 \text{ rad per S per V}$$

Then the value of K_o follows:

$$K_o = \frac{11.2 \, (10^6)}{s} \text{ rad per S per V}$$

The s in the denominator converts the frequency characteristics of the VCM to phase; i.e., phase is the integral of frequency. The gain constant for the MC4044/4344 phase detector is found on its data sheet as $K_p = 0.111$ V per rad.

Because a type 2 system is required (phase coherent output), the loop transfer function must take the form of Equation 8-17. The parameters thus far determined include K_p, K_o, K_n, leaving only K_f as the variable for design. Writing the loop transfer function and relating it to Equation 8-17 yields the following equation:

$$G(s)H(s) = \frac{K_p K_v K_n K_f}{s} = \frac{K(s + a)}{s^2} \qquad \textbf{Equation 8-34}$$

Hence, the value of K_f must take the form as follows:

$$K_f = \frac{s + a}{s} \qquad \textbf{Equation 8-35}$$

The circuit shown in Fig. 8-10 will simulate the desired result for K_f in the form expressed by Equation 8-36:

$$K_f = \frac{R_2 C s + 1}{R_1 C s} \qquad \textbf{Equation 8-36}$$

The parameters R_1, R_2, and C are the variables employed to establish the overall loop characteristics.

The MC4044/4344 provides the active circuitry required to simulate the filter K_f. An additional low-current, high-beta buffering device or FET can be used to boost the input impedance to minimize the leakage current from capacitor C between sample updates. As a result, longer sample periods are achievable.

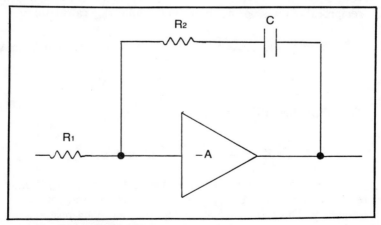

Fig. 8-10. Active filter design.

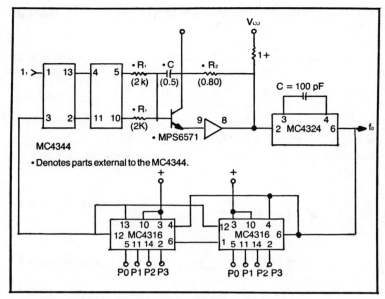

Fig. 8-11. Circuit diagram of a type 2 phase-locked loop.

Because the gain of the active filter circuitry in the MC4044/4344 is not infinite, a gain correction factor, K_c, must be applied to K_f in order to properly characterize the function. The value of K_c was found experimentally to be equal to 0.5. Therefore, you can define a new value K_{fc} as shown by Equation 8-37.

$$K_{fc} = K_f K_c = 0.5 \left[\frac{R_2 C s + 1}{R_1 C s} \right]$$ **Equation 8-37**

Note that Equation 8-37 only applies for the MC4044/4344. When the voltage gain of the active device is infinite, Equation 8-36 applies.

The PLL circuit diagram is shown in Fig. 8-11. Its Laplace representation is shown in Fig. 8-12. The following equations can be written from Fig. 8-12. The loop transfer function is written as follows:

$$G(s)H(s) = \qquad K_p K_{fc} K_o K_n \qquad \text{**Equation 8-38**}$$

$$= K_p (0.5) \; \frac{R_2 C s + 1}{R_1 C s} \; \frac{K_v}{s} \; \frac{1}{N} \qquad \text{**Equation 8-39**}$$

The characteristic equation takes the form shown in Equation 8-40.

$$C.E. = 1 + G(s)H(s) = 0 \qquad \text{Equation 8-40}$$

$$= s^2 + \frac{0.5 \, K_p K_v R_2}{R_1 N} s + \frac{0.5 \, K_p K_v}{R_1 CN} \qquad \text{Equation 8-41}$$

Relating Equation 8-41 to the normalized standard form shown in Equation 8-26, and equating like coefficients yields the following results:

$$\omega^2_n = \frac{0.5 \, K_p K_v}{R_1 CN} \qquad \text{Equation 8-42}$$

$$2 \zeta \omega_n = \frac{0.5 \, K_p K_v R_2}{R_1 N} \qquad \text{Equation 8-43}$$

With the use of an active filter whose open-loop gain (A) is very large, $K_c = 1$ and Equations 8-42 and 8-43 have the following form:

$$\omega^2_n = \frac{K_p K_v}{R_1 CN} \qquad \text{Equation 8-44}$$

$$2 \zeta \omega_n = \frac{K_p K_v R_2}{R_1 N} \qquad \text{Equation 8-45}$$

The percent overshoot and settling time are next used to determine ω_n. Figure 8-7 that a damping ratio of 0.8 will produce a peak overshoot less than 20 percent and will settle to within 5 percent at $\omega_n t = 4.5$. The required lockup time is 1 mS. Then $\omega_n = 4.5/t = 4.5/0.001 = 4500$ rad per S. Solving for $R_1 C$ from Equation 8-42, you have the following equation:

$$R_1 C = \frac{0.5 \, K_p K_v}{\omega^2_n N} \qquad \text{Equation 8-46}$$

Substitution of the known quantities into Equation 8-46 yields the following results:

$$R_1 C = \frac{(0.5)\,(0.111)\,(11.2)\,(10^6)}{(4500)^2\,(30)}$$

$$= 0.00102$$

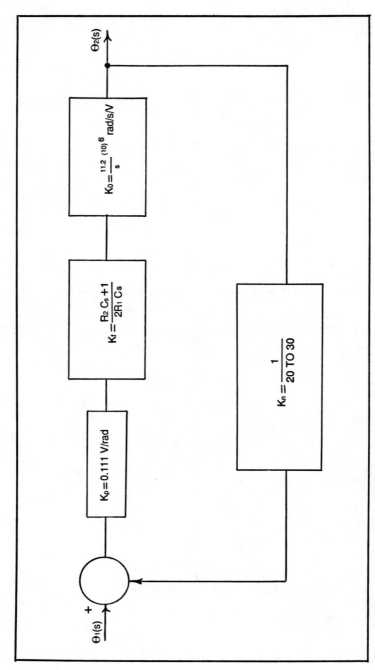

Fig. 8-12. Laplace representation of the circuit shown in Fig. 8-11.

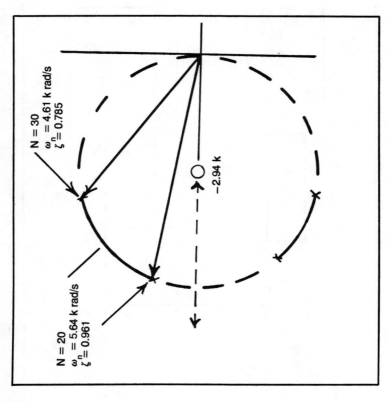

Fig. 8-13. Root locus variation.

Remember that maximum overshoot occurs at N_{max}, which is minimum loop gain. Select $C = 0.5\ \mu F$. Then calculate the value for R_1 employing the results of the preceding equation.

$$R_1 = \frac{0.00102}{C} = \frac{0.00102}{5(10^{-6})}$$

$$= 2000\ \text{ohms}$$

Solving for R_2 in Equation 8-43 yields the following result:

$$R_2 = \frac{2\,\zeta\,\omega_n\,R_1\,N}{K_p\,K_v\,(0.5)} = \frac{2\,\zeta}{C\,\omega_n} \qquad \textbf{Equation 8-47}$$

$$R_2 = \frac{2(0.8)}{5(10^{-7})\,(4500)} = 711\ \text{ohms}$$

Select R_2 to be equal to 680 ohms, a standard value of resistance.

All circuit parameters have been determined and the PLL can be properly configured. The loop gain is a function of the divide ratio, K_n, so the closed-loop poles will vary in position as K_n varies. The root locus in Fig. 8-13 shows the closed-loop pole variation.

The loop was designed for the programmable counter, N = 30. The system response for N = 20 exhibits a wider bandwidth and larger damping factor, thus reducing both lockup time and percent overshoot.

The type 2, second-order loop was illustrated as a design example because it provides excellent performance for both type 1 and type 2 applications. Even in systems that do not require phase coherency, a type 2 loop still offers an optimum design.

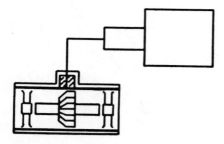

Glossary

amplitude ratio: The size relationship between two quantities obtained by dividing one size by another. Specifically, the ratio of peak height of an output signal to the peak height of related input signa. See also *magnitude ratio* and *gain*.

attenuation: Production of an output signal of smaller size than the corresponding input, sometimes termed a gain of less than 1.0 or an amplitude ratio of less than 1.0. For input waves of a specific frequency, the attentuation is the ratio of peak height of output wave to peak height of input wave when that ratio is less than 1.0. See *gain*.

automatic reset: See *reset action*.

block: A part of a control loop which receives an input signal and transmits an output signal. It makes the output signal differ from the input by either changing the size, delaying the signal, or altering the way in which the size changes with time.

Bode plot: A graph showing how an output signal differs from its input signal in size and timing, when the input is a wavelike signal of specific frequency. It shows for each frequency the magnitude ratio and phase angle between output signal and input signal. It is plotted to show how magnitude ratio and phase angle change with the frequency of the input signal. See *Nyquist plot*.

break frequency: Another name for *corner frequency*.

build-up time: Lenth of time it takes the measured variable to first reach a new setpoint when a small change in setpoint is made with the controller tuned to overshoot and cycle into the new value. This term is also called *rise time*.

capacitance: The change in energy or material required to make a unit change in a measured variable, such as BTU per degree of temperature rise, or cubic feet of contents per foot of increase in level. For an integrating block, the reciprocal or capacitance and a known ratio of flow can be used to compute the gain per minute.

capacity: Maximum quantity of energy or material that can be stored within the confines of a stated piece of equipment, such as volume of liquid a tank will hold when full. Different from *capacitance*.

capacity lag: See *transfer lag*.

cascade control system: A control system where the output of one controller is used to adjust the setpoint of second controller, and the output of the second controller actually adjusts the control element.

closed loop: The complete signal path in a control system, represented as a group of units connected to a process in such a manner that a signal started at any point follows a closed path and comes back to that point. The loop includes the deviation detecting point as well as the control action and process equipment.

control action: The kind of correction the controller makes for a deviation proportional action, reset action, and rate action. See *control mode*.

control element: The final controlling means, such as a control valve to change flow, pressure, or level, or a heater adjusting device to change temperature.

control loop: A path along which control affecting signals travel from the deviation detecting point, through the controller, through the process equipment, and back to the deviation detecting point. See *closed loop* and *open loop*.

control mode: One of the three principal kinds of control action, proportional, reset, and rate.

control point: The value of the controlled variable when it has lined out after a disturbance. This is not always the same as the setpoint.

corner frequency: On a bode plot, the frequency where the magnitude ratio begins to decrease, located by the intersection of the zero line and the asymptote of the downward slope. It is a measure of the speed of response of the block to an incoming signal. The higher the corner frequency is, the faster the response is.

critical damping: See *damping*.

damping: A measure of something like the restraining force which prevents return to an equilibrium position. Applies to response like that of a pendulum with friction at the suspension point or a suspended spring with a weight attached and a viscous fluid around the weight. The *damping factor* measures the amount of damping. With critical damping, return to equilibrium is as fast as possible without overshooting. The critical damping factor is 1.0. With overdamping, the restraining force is greater, and return is slower than with critical damping. The damping factor is greater than 1.0 for overdamping. With underdamping, the return to equilibrium is faster, and the equilibrium point is overshot and reached only after some cycling around it. The damping factor is less than 1.0 for underdamping.

damping factor: A number evaluating the amount of damping. The symbol for damping factor is called zeta (ζ). See *damping*.

dead time: A period of delay between two related actions, such as the beginning of a change in an input signal and the beginning of a related change in output. Also called *transportation lag* and *distance velocity lag*.

decade: A frequency span whose upper limit is ten times its lower limit.

decibel: A measure of magnitude ratio used for convenience; that is, decibels equals 20 times the common logarithm of the magnitude ratio. The symbol for decibel is dB.

decibel curve: On a Bode plot, the curve of signal size relationship (output to input) against signal frequency, the size ratio being expressed in dB. See *decibel* and *magnitude curve*.

deviation: The difference between the setpoint and the value of the measured variable, usually in percent of full scale span of the variable. It is negative when the measured value is below the setpoint. See *error*.

deviation ratio: The factor by which automatic control reduces the deviations resulting from sinusoidal disturbances in a process. For each frequency of disturbance, it compares the values of deviation with and without the controller. A ratio of over 1.0 means the deviation is larger with the controller. See *error ratio*.

deviation reduction factor: Like deviation ratio, this is a factor evaluating how much the controller reduces the deviations caused by disturbances. The higher the deviation reduction factor is the smaller the controlled deviations are. The value depends on the form and point of occurrence of the disturbance.

differential gap: A neutral zone in contact actuating forms of control. A small range of values near the setpoint where no change in controller output will occur.

distance velocity lag: See *dead time*.

disturbance: A change in a condition outside the loop which affects the control signal.

dynamic analysis: Study of control system performance at times of disturbance in the controlled variable or in conditions which affect that variable.

error: The term used instead of *deviation*.

error ratio: See *deviation ratio*.

feedback control system: A control loop in which the measured variable is fed back and compared with a desired value, and the difference is used as the input to the controller.

final control element: See *control element*.

final curves: The Bode plot of the control system.

first-order lag: A term used to describe the signal delaying and signal size changing effects of a part of the control loop. The name comes from the form of the equation which represents the relation between output and input. The effects are evaluated by a time constant.

follower operation: Control operation with the load fixed and setpoint variable. In other words, the follower acts to make the variable follow the setpoint.

frequency: The number of vibrations or waves in a unit of time.

frequency response: The relationship between output signal and input signal of a system or component, when the input is a sinusoidal wave and the relationship is given for the complete range of frequencies.

frequency response tests: Tests of the output of a part of a control system to show how it changes when the input is a cycling signal, the cycles having a specific frequency and tests being made with many specific frequencies.

which has a value of less than 1.0 is actually a signal loss and

gain: The following meanings which are related to signal size: 1. The ratio of the output signal peak height to input signal peak height for an input wave of specific frequency. This is the same as magnitude ratio for frequency response. 2. The ratio of change in output of part or all of a control system to its change in input, each value of output being measured for a given value of input. 3. For blocks whose output will line out for a given value of input, gain may be expressed in either of two forms: a. Units of output per unit of input. b. Ratio of the percentage difference in output to the percentage difference in input. In either form, this gain is also called steady state gain. See *steady state*. Steady state gain is also called zero frequency gain or static gain. A steady state gain in form 3b

may be called either attentuation or gain. For blocks whose output does not line out for a given positive value of input, but does stabilize by reaching a constant rate of change, the steady state gain is expressed as a gain per unit time, the amount of gain being expressed either in units per unit time as in 3a, or in percentage ratio per unit time as in 3b.

gain margin: A measure of the margin of stability of a control system, based on the signal size ratio. On a frequency response plot, the negative of the dB value of the final magnitude curve at the −180-degree phase frequency. For example, if the magnitude curve is at −15 dB at that frequency, the gain margin is +15 dB. In other words, when this value is written as a magnitude ratio rather than in dB, it is 1 minus the magnitude ratio at the −180-degree frequency, rather than the factor which must be multiplied by the magnitude ratio to make the product equal to 1.0.

gain steady state: See *gain*.

input: A condition which adds energy or material to the control system. The input can be a manipulated variable, disturbance, or the reference input.

input signal: The control signal when it enters one of the blocks or summing point. See *input.*

integral control: Same as reset action.

integral time: In reset action of a controller, the reciprocal of repeats per minute.

integrating block: A block whose output increases steadily when the input is a constant positive value. This block totals the product of input and time and provides a continuous measure of the area under the curve of input versus time. Mathematically, it takes the integral of the input.

lag: A block which delays or smooths an input signal. It might change the size or timing of the signal, the timing effect sometimes changing the shape of the signal. See *first-order lag, second-order lag,* and *dead time.*

lead: On a Bode plot, a response represented by inverting the phase and magnitude curves for a lag. It acts to advance the output signal with respect to the input, in contrast to the retarding action of a lag.

load: The conditions which determine the amount of energy or material that must be supplied to a process to maintain the variable at the desired level. A change in load results in use of a different amount of material or energy to produce the same value of the variable.

magnitude curve: On a Bode plot, the curve of signal size ratio (output to input) against signal frequency, the size ratio being expressed either in dB or in magnitude ratio.

magnitude ratio: The ratio of the peak size of a cycling output signal to the peak size of the cycling input signal. Also called the *amplitude ratio.* When the ratio values are plotted against the input signal frequency, the plotted curve may be called the *magnitude curve.*

manipulated variable: That quantity or condition which the control elements apply to the controlled system. This is the process input over which you have control.

measured variable: The process condition (such as temperature or pressure) selected to represent the state of material which is being made or processed. See *process.*

measuring error: The difference between the measured value of the controlled variable and the true value. See *error* and *deviation*.

natural frequency: The frequency of the cycling motion of an undamped second-order component. For a transient response curve, the frequency of cycling which the deviation would have if the response were undamped. Also, the frequency at which an object would vibrate at zero damping. On a Bode plot of a second-order lag, the frequency where the extension of the final slope of the dB curve intersects the zero dB line.

Nichols chart: A chart used for plotting either the open-loop or closed-loop response of a control system, and converting one to the other.

nonlinearity: Dependence of the output signal response not strictly upon the input signal but upon particular signal input conditions or upon operating peculiarities of the device. The name signifies that the relation between output and input is not representable by a single straight line.

Nyquist plot: A chart of magnitude and phase angle of output compared to input, plotted on polar coordinates, to show how magnitude and phase angle change with the frequency of the signal. The magnitude becomes an arrow length from the central point; the phase, the angle the arrow makes with a zero line; the plot is a line connecting the arrow points.

octave: A frequency span whose upper limit is twice its lower limit.

offset: The difference between the value of the measured variable and the setpoint when the control system is at a steady state.

open loop: All of the control loop except the point where the measured value is compared with the setpoint to give the deviation. The open loop response refers to the operation of a feedback control system with the feedback loop disconnected at the deviation detecting point.

output (other than output signal): For a process element or equipment in the control loop, an outflow which takes material or energy away from the equipment, like heat loss or liquid flow.

output signal: The control loop signal when it leaves any part of the loop. Sometimes simply called *output*.

overdamping: See *damping.*

peak frequency: On a Bode plot, the frequency of a high point of the magnitude curve before the curve drops downward.

phase angle: A measure of the time by which a cycling output lags behind a cycling input, measured in degrees as a part of the full 360 degrees of one cycle. Also, the angle by which a sinusoidal output differs from a sinusoidal input. The angle is considered to be negative when the output lags behind the input and positive when the output is advanced with respect to the input. Sometimes called phase shift, defined as the angular difference between corresponding points on input and output signal wave shapes, disregarding any difference in size of signal. Sometimes simply called phase.

phase curve: On a Bode plot, the curve showing the time relationship between output and input signals as measured by the frequency response method in terms of phase angle.

phase lag: The angle by which the cycling output lags behind a sinusoidal input. Sometimes called phase. It is a negative phase angle if there is a phase lag of 30 degrees or a phase angle of -30 degrees.

phase lead: The angle by which the cycling output is advanced with respect to a sinusoidal input. A phase lead of 30 degrees is a phase angle of $+30$ degrees.

phase margin: On a frequency response plot, the angular difference between -180 degrees and the phase angle at the frequency of 0 dB. For example, if the phase angle is -135 degrees where the magnitude is 0 dB, the phase margin is $+45$ degrees. It is a measure of the margin of stability of the control system. The frequency response plot it then applies to is that of the open loop, including the control actions.

phase shift: See *phase angle.*

plant: Sometimes means all of the process equipment, and occasionally used interchangeably with process but generally distinguished from the process by applying to hardware rather than to energy and material relationships.

process: The manufacturing operations which use energy measurable by some quantity such as temperature, pressure, or flow to produce changes in quality or quantity of some material or energy.

process variable: see *measured variable.*

proportional band: The percent of the range of the measured variable for which a proprotional controller will produce a 100 percent range in its output. The percent of range of the measured variable corresponding to a 100 percent change in position of the control element.

proportional control: The control action producing a controller output proportional to the size of the deviation.

quadratic: Another name for second order, based on the form of equation which represents the second-order response.

radian: An angle whose arc is equal to the radius of the circle in which the angle is measured. There are 2π radians in 360 degrees, or one radian is about 57.3 degrees.

radian frequency: This term refers to frequency multiplied by 26π, which is equal to ω (omega) the angular velocity. Many times in control systems the radian frequency is simply referred to as freqeuncy.

ramp signal: A signal which is changing linearly with time; that is, a straight line increase or decrease in value.

rate action: A control action which affects the controller output whenever the deviation is changing, and affects it more when the deviation is changing faster. It may either increase or decrease the controller output, depending on whether the deviation is increasing or decreasing and whether it is negative or positive.

rate amplitude: The ratio of the maximum output of proportional plus rate action to the steady output of proportional action alone, after a step increase in deviation, both output being measured from the same baseline, and the controller output before the step.

rate time: The difference in time taken by proportional action alone and by proportional plus rate to produce the same controller output for a ramp change in deviation.

rate time constant: Here, the time it takes 63.2 percent of the rate action to disappear after the deviation stops changing. Its value is rate time divided by *rate amplitude*.

ratio control system: A control system that maintains two or more variables at a predetermined ratio by making the value of one variable adjust the controller setpoint for another variable.

reference input: Sometimes used in control system descriptions to mean the *setpoint*. Often used in feedback control to mean the desired value.

regulator operation: Control operation with the setpoint fixed and the load variable. The regulator acts to keep the measured variable constant in spite of disturbances or load changes.

repeats per minute: The reset adjustment in repeats per minute is the number of times per minute that the proportional action is repeated by reset. It is the reciprocal of the integral time of the reset action.

reset action: A control action which adjusts the controller output in accordance with both the size of the deviation and the time it lasts. Sometimes called integral action because it affects the controller output by the integral of the deviation.

reset rate: Same as *repeats per minute*.

resistance: Opposition to the flow of heat, current, or fluids. It is sometimes a measure of the change in flow of energy or material.

rise time: Length of time it takes the measured variable to change 63.2 percent of the way to its new value for a new setpoint when a small change in setpoint is made.

second-order lag: A term used to describe the signal delaying and signal size changing effects of a part of the control loop. The name comes from the form of the equation which represents the relationship between output and input. The effects are evaluated by a damping factor and natural frequency.

self-regulating process: One in which the output will line out for each value of input. An integration block is not self-regulating.

sensitivity: The ratio of controller output response to a specified change in the measured variable. It is always expressed in units of output per unit of input, such as psi per °F or psi per inch of pen motion. This term is often used to describe the output input relation of part or all of the control loop to identify an output as sensitive or insensitive to changes in input.

servo operation: Same as *regulator operation*.

setpoint: The desired value of the variable which is being measured and controlled. Also called the *reference input*.

settling time: The time which the variable will take to line out within 5 percent of a new setpoint after a step change.

signal: Information being transmitted from one part of the control system to another by some change in energy or material.

signature curve: For a process, the transient response curve of the variable after a step change in the position of the final control element. This curve is obtained with the controller disconnected and shows the response of the process as part of the control loop. Also called the process signature curve and the process reaction.

sine wave: A regular, fluctuating signal varying periodically with time. The wave varies as the sine of the angle made by a line rotating around a point at a constant speed.

static gain: See *gain*.

steady state: That condition when the input signal is a constant value and the output has stabilized either by leveling out at a constant value or by reaching a constant rate of change. Also that condition when the input signal is a constant amplitude cycle, such as a sine wave, and the output signal has also become a constant amplitude cycle.

steady state gain: See *gain*.

step change: An input change of definite value made in a very short, like a sudden change in setpoint. The values are constant before and after the step.

three capacity process: A process which can be represented in block diagram form by a second-order lag plus a first-order lag, or by three first-order lags.

time constant: Time required for the output of a first-order lag device to reach 63.2 percent of its final value for a step change in input. The output will reach 95 percent in three time constants, 98 percent in four, and over 99 percent in five time constants.

total curves: The Bode plot of all of the control system except control actions.

transfer function: An expression stating the relationship between an input signal and a corresponding output signal, the relationship involving both the size and timing of the signal. Sometimes defined as the ratio of Laplace transform of output and input signal.

transfer lag: Any lag except dead time. Sometimes called capacity lag.

transient response: How the output of blocks or control systems changes with time after a disturbance.

transportation lag: Same as *distance/Velocity lag dead time*.

two capacity process: A process which can be represented in block diagram form as one second-order lag or two first-order lags.

ultimate period: The cycle time for deviation cycles shown when a proportional controller is adjusted for the ultimate proportional band. Also called the ultimate cycle.

ultimate proportional band: The proportional band which produces a continuously cycling deviation of constant peak magnitude when the control action is proportional only and a small disturbance occurs.

underdamping: See *damping*.

variable: Same as *measured variable*.

zero frequency gain: See *gain*.

Appendix A
Laplace Transform Table

No.	F(s)	f (t) ; t > 0
1.	1	$\delta(t)$ unit impulse
2.	e^{-Ts}	$\delta(t - T)$ delayed impulse
3.	$\dfrac{1}{s + a}$	e^{-at}
4.	$\dfrac{1}{(s + a)^\eta}$	$\dfrac{1}{(\eta - 1)!} \; t^{\eta-1} e^{-at}$ $\eta = 1, 2, 3, \ldots$
5.	$\dfrac{1}{(s + a)(s + b)}$	$\dfrac{1}{b - a} \; (e^{-at} - e^{-bt})$
6.	$\dfrac{s}{(s + a)(s + b)}$	$\dfrac{1}{a - b} \; (ae^{-at} - be^{-bt})$
7.	$\dfrac{s + z}{(s + a)(s + b)}$	$\dfrac{1}{b - a} \; [(z - a)e^{-at} - (z - b)e^{-bt}]$
8.	$\dfrac{1}{(s + a)(s + b)(s + c)}$	$\dfrac{e^{-at}}{(b - a)(c - a)} + \dfrac{e^{-bt}}{(c - b)(a - b)} + \dfrac{e^{-ct}}{(a - c)(b - c)}$
9.	$\dfrac{s + z}{(s + a)(s + b)(s + c)}$	$\dfrac{(z - a)e^{-at}}{(b - a)(c - a)} + \dfrac{(z - b)e^{-bt}}{(c - b)(a - b)} + \dfrac{(z - c)e^{-ct}}{(a - c)(b - c)}$
10.	$\dfrac{\omega}{s^2 + \omega^2}$	$\sin \omega t$
11.	$\dfrac{s}{s^2 + \omega^2}$	$\cos \omega t$
12.	$\dfrac{s + z}{s^2 + \omega^2}$	$\sqrt{\dfrac{z^2 + \omega^2}{\omega^2}} \; \sin(\omega t + \phi)$ $\phi \equiv \tan^{-1}(\omega/z)$
13.	$\dfrac{s \sin \phi + \omega \cos \phi}{s^2 + \omega^2}$	$\sin(\omega t + \phi)$
14.	$\dfrac{1}{(s + a)^2 + \omega^2}$	$\dfrac{1}{\omega} \; e^{-at} \sin \omega t$

No.	F(s)	f(t); t > 0
15.	$\dfrac{1}{s^2 + 2\zeta\omega_\eta s + \omega^2{}_\eta}$	$\dfrac{1}{\omega_d}\, e^{-\zeta\omega_\eta t} \sin \omega_d t \qquad \omega_d \equiv \omega_\eta \sqrt{1 - \zeta^2}$
16.	$\dfrac{s + a}{(s + a)^2 + \omega^2}$	$e^{-at} \cos \omega t$
17.	$\dfrac{s + z}{(s + a)^2 + \omega^2}$	$\sqrt{\dfrac{(z - a)^2 + \omega^2}{\omega^2}}\, e^{-at} \sin (\omega t + \phi) \quad \phi \equiv \tan^{-1}\left(\dfrac{\omega}{z - a}\right)$
18.	$\dfrac{1}{s}$	$\mu(t)$ or 1 unit step
19.	$\dfrac{1}{s}\, e^{-Ts}$	$\mu(t - T)$ delayed step
20.	$\dfrac{1}{s}\, (1 - e^{-Ts})$	$\mu(t) - \mu(t - T)$ rectangular pulse
21.	$\dfrac{1}{s(s + a)}$	$\dfrac{1}{a}\, (1 - e^{-at})$
22.	$\dfrac{1}{s(s + a)(s + b)}$	$\dfrac{1}{ab}\, (1 - \dfrac{be^{-at}}{b - a} + \dfrac{ae^{-bt}}{b - a})$
23.	$\dfrac{s + z}{s(s + a)(s + b)}$	$\dfrac{1}{ab}\, (z - \dfrac{b(z - a)e^{-at}}{b - a} + \dfrac{a(z - b)e^{-bt}}{b - a})$
24.	$\dfrac{1}{s(s^2 + \omega^2)}$	$\dfrac{1}{\omega^2}\, (1 - \cos \omega t)$
25.	$\dfrac{s + z}{s(s^2 + \omega^2)}$	$\dfrac{z}{\omega^2} - \sqrt{\dfrac{z^2 + \omega^2}{\omega^4}}\, \cos (\omega t + \phi) \qquad \phi \equiv \tan^{-1}(\omega/z)$
26.	$\dfrac{1}{s(s^2 + 2\zeta\omega_\eta s + \omega^2\eta)}$	$\dfrac{1}{\omega^2{}_\eta} - \dfrac{1}{\omega_\eta \omega_d}\, e^{-\zeta\omega_\eta t} \sin (\omega_d t + \phi)$ $\omega_d \equiv \omega_\eta \sqrt{1 - \zeta^2} \qquad \phi \equiv \cos^{-1}\zeta$
27.	$\dfrac{1}{s(s + a)^2}$	$\dfrac{1}{a^2}\, (1 - e^{-at} - ate^{-at})$
28.	$\dfrac{s + z}{s(s + a)^2}$	$\dfrac{1}{a^2}\, [z - ze^{-at} + a(a - z)te^{-at}]$
29.	$\dfrac{1}{s^2}$	t unit ramp
30.	$\dfrac{1}{s^2(s + a)}$	$\dfrac{1}{a^2}\, (at - 1 + e^{-at})$
31.	$\dfrac{1}{s^\eta}\ \eta = 1, 2, 3, \dots$	$\dfrac{t^{\eta-1}}{(\eta - 1)!} \qquad 0! = 1$
32.	$\dfrac{1 + as}{s^2(1 + ts)}$	$t + (a - T)(1 - e^{-t/T})$

No.	F(s)	f (t); t > 0
33.	$\dfrac{s}{s^2 - \omega^2}$	$\cosh \omega t$
34.	$\dfrac{\omega}{s^2 - \omega^2}$	$\sinh \omega t$
35.	$\dfrac{s^2 - 2}{s(s^2 - 4)}$	$\cosh^2 t$
36.	$\dfrac{2}{s(s^2 - 4)}$	$\sinh^2 t$
37.	$\dfrac{2s\omega}{(s^2 + \omega^2)^2}$	$t \sin \omega t$
38.	$\dfrac{s^2 - \omega^2}{(s^2 + \omega^2)^2}$	$t \cos \omega t$
39.	$\dfrac{\eta}{2} \left[\dfrac{1}{(s - \omega)^{\eta+1}} + \dfrac{1}{(s + \omega)^{\eta+1}} \right]$	$t^\eta \cosh \omega t$
40.	$\dfrac{\eta!}{2} \left[\dfrac{1}{(s - \omega)^{\eta+1}} - \dfrac{1}{(s + \omega)^{\eta+1}} \right]$	$t^\eta \sinh \omega t$

No.	F(s)	f(t); t > 0
41.	$\dfrac{2\,\omega^3}{s^4 - \omega^4}$	$\sinh \omega t - \sin \omega t$
42.	$\dfrac{2\,\omega^2 s}{s^4 - \omega^4}$	$\cosh \omega t - \cos \omega t$
43.	$\dfrac{2\,\omega s^2}{s^4 - \omega^4}$	$\sinh \omega t + \sin \omega t$
44.	$\dfrac{2 s^3}{s^4 - \omega^4}$	$\cosh \omega t + \cos \omega t$
45.	$\dfrac{1}{s}\ (e^{-as} - e^{-bs})$	$\mu(t-a) - \mu(t-b)$
46.	$\dfrac{\omega}{s^2 + \omega^2}\ (1 + e^{\frac{-\pi s}{\omega}})$	$[\mu(t) - \mu(t- \frac{\pi}{\omega})]\sin \omega t$
47.	$\dfrac{1}{as^2} - \dfrac{e^{-as}}{s(2 - e^{-as})}$	$[\mu(t) - \mu(t-a)]\ \dfrac{t}{a}$ SAW TOOTH
48.	$\dfrac{1}{s}(1 - e^{-as})^2 = \dfrac{1}{s}\tanh\dfrac{as}{2}$	$\mu(t) - 2\mu(t-a) + \mu(t-2a)$ $\pm$ etc. SQUARE WAVE

No.	F(s)	f (t) ; t > 0
49.	$\dfrac{(1 - e^{-as})^2}{as^2} = \dfrac{1}{as^2} \tanh \dfrac{as}{2}$	$\dfrac{1}{a} [\mu(t) t - 2\mu(t-a) (t-a) \\ t\mu(t-2a) (t-2a)]$ TRIANGULAR WAVE
50.	$\left(\dfrac{\omega}{s^2 + \omega^2}\right) \left[\dfrac{1}{1 - e^{-\frac{\pi s}{\omega}}}\right]$	$\mu(t) \sin\omega t + \mu(t-\pi/\omega) \sin \omega(t-\pi/\omega)$ HALF WAVE RECTIFIED SINE WAVE
51.	$\left(\dfrac{\omega}{s^2 + \omega^2}\right) [1 + e^{-\frac{\pi s}{\omega}}] =$ $\dfrac{\omega}{(s^2 + \omega^2)} \left[\coth \dfrac{\pi s}{2\omega}\right]$	$\mu(t) \sin \omega t + \mu(t- \dfrac{\pi}{\omega}) \sin \omega (t- \dfrac{\pi}{\omega})$ FULL WAVE RECTIFIED SINE WAVE

Appendix B
Laplace Operations Table

No.	F(s)	F(t); $t > 0$
1.	$F(s) = \int_0^\infty f(t)\, e^{-st}\, dt$	$f(t)$
2.	$F(s - a)$	$e^{at}\, F(t)$
3.	$\dfrac{1}{a}\, F\left(\dfrac{s}{a}\right)$	$F(at)$
4.	$s\, F(s) - f(o)$	$\dfrac{Df(t)}{dt} = f'(t)$
5.	$s^n F(s) - s^{n-1} f(0) - s^{n-2} f'(0)$ $- \ldots - f^{n-1}(0)$	$\dfrac{d^n f(t)}{dt^n} = f^n(t)$
6.	$\dfrac{1}{s}\left[F(s) + f^{-1}(0)\right]$	$\int f(t)\, dt = f^{-1}(t)$
7.	$\dfrac{F(s)}{s^n} + \dfrac{f^{-1}(0)}{s^n} + \dfrac{f^{-2}(0)}{s^{n-1}}$ $+ \ldots + \dfrac{f^{-n}(0)}{s}$	$\int\int \ldots \int f(t) dt^n = f^{-n}(t)$
8.	$e^{-as} F(s)$	$\mu(t - a)\, f(t - a)$
9.	$F(s)\, G(s)$	$\int_0^t f(\tau)\, g(t - \tau)\, d\tau$

Appendix C
MC1741/MC1741C
Operational Amplifiers

INTERNALLY COMPENSATED, HIGH
PERFORMANCE MONOLITHIC OPERATIONAL AMPLIFIER

...designed for use as a summing amplifier, integrator, or amplifier
with operating characteristics as a function of the external feedback
components.

- [ ] No Frequency Compensation Required
- [ ] Short-Circuit Protection
- [ ] Offset Voltage Null Capability
- [ ] Wide Common-Mode and Differential Voltage Ranges
- [ ] Low-Power Consumption
- [ ] No Latch Up
- [ ] Same Pin Configuration as the MC1709

Table C-1. Maximum Ratings (TA = + 25° C Unless Otherwise Noted.)

Rating	Symbol	MC1741C	MC1741	Unit
Power Supply Voltage	V_{CC} V_{EE}	+18 -18	+22 -22	Vdc Vdc
Input Differential Voltage	V_{ID}	±30		Volts
Input Common Mode Voltage (Note 1)	V_{ICM}	±15		Volts
Output Short Circuit Duration (Note 2)	t_S	Continuous		
Operating Ambient Temperature Range	T_A	0 to +70	-55 to +125	°C

Storage Temperature Range
  Metal, Flat and Ceramic Packages
  Plastic Packages
	T_{stg}	-65 to +150
		-55 to +125
				°C

Junction Temperature Range
  Metal and Ceramic Packages
  Plastic Packages
	T_J	175
		150
				°C

Note 1. For supply voltages less than ± 15 V, the absolute maximum input voltage is equal to the supply voltage.

Note 2. Supply voltage equal to or less than 15 V.

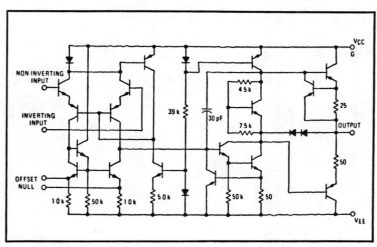

Fig. C-1. Equivalent circuit schematic.

Table C-2. Thermal Information. (continued on page 297)

(continued on page 297)

The maximum power consumption an integrated circuit can tolerate at a given operating ambient temperature, can be found from the equation:

$$PD_{(TA)} = \frac{T_{J(max)} - T_A}{R_{\theta JA}(Typ)}$$

Where: $PD_{(TA)}$ = Power Dissipation allowable at a given operating ambient temperature. This must be greater than the sum of the products of the supply voltages and supply currents at the worst case operating condition.

$T_{J(max)}$ = Maximum Operating Junction Temperature as listed in the Maximum Ratings Section

T_A = Maximum Desired Operating Ambient Temperature

$R_{\theta JA}(Typ)$ = Typical Thermal Resistance Junction to Ambient

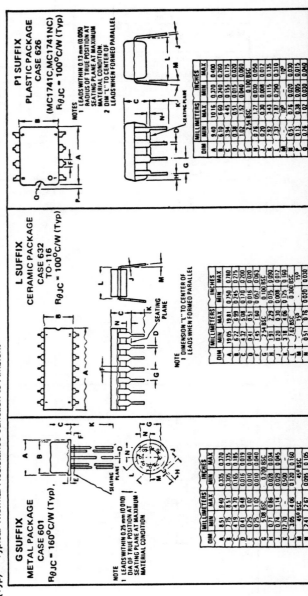

Table C-2. Thermal Information. (continued from page 296).

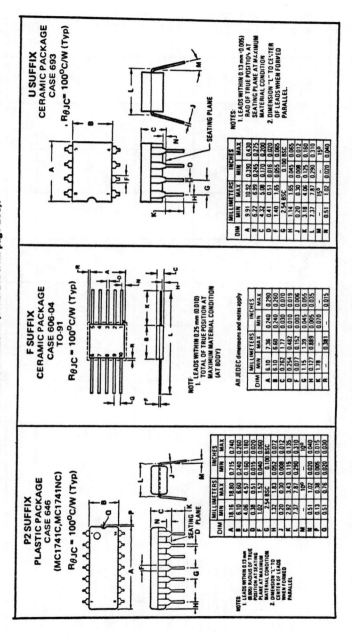

P2 SUFFIX
PLASTIC PACKAGE
CASE 646
(MC1741C, MC1741NC)
$R_{\theta JC} = 100°C/W$ (Typ)

DIM	MILLIMETERS		INCHES	
	MIN	MAX	MIN	MAX
A	18.16	18.80	0.715	0.740
B	6.10	6.60	0.240	0.260
C	4.06	4.57	0.160	0.180
D	0.38	0.51	0.015	0.020
F	1.02	1.52	0.040	0.060
G	2.54 BSC		0.100 BSC	
H	1.32	1.83	0.052	0.072
J	0.20	0.30	0.008	0.012
K	2.92	3.43	0.115	0.135
L	7.37	7.87	0.290	0.310
M	–	10°	–	10°
N	0.51	1.02	0.020	0.040
P	0.13	0.38	0.005	0.015
Q	0.51	0.76	0.020	0.030

NOTES:
1. LEADS WITHIN 0.13 mm (0.005) RADIUS OF TRUE POSITION AT SEATING PLANE AT MAXIMUM MATERIAL CONDITION.
2. DIMENSION "L" TO CENTER OF LEADS WHEN FORMED PARALLEL.

F SUFFIX
CERAMIC PACKAGE
CASE 606-04
TO-91
$R_{\theta JC} = 100°C/W$ (Typ)

NOTE: LEADS WITHIN 0.25 mm (0.010) TOTAL OF TRUE POSITION AT MAXIMUM MATERIAL CONDITION (AT BODY)

All JEDEC dimensions and notes apply

DIM	MILLIMETERS		INCHES	
	MIN	MAX	MIN	MAX
A	6.10	7.36	0.240	0.290
B	6.10	6.60	0.240	0.260
C	0.762	1.77	0.030	0.070
D	0.254	0.482	0.010	0.019
F	0.077	0.152	0.003	0.006
G	1.15	1.39	0.045	0.055
H	0.127	0.889	0.005	0.035
K	1.78	–	0.070	–
R	–	0.381	–	0.015

U SUFFIX
CERAMIC PACKAGE
CASE 693
$R_{\theta JC} = 100°C/W$ (Typ)

DIM	MILLIMETERS		INCHES	
	MIN	MAX	MIN	MAX
A	9.91	10.92	0.390	0.430
B	6.22	6.99	0.245	0.275
C	4.32	5.08	0.170	0.200
D	0.41	0.51	0.016	0.020
F	1.40	1.65	0.055	0.065
G	2.54 BSC		0.100 BSC	
H	1.14	1.65	0.045	0.065
J	0.20	0.30	0.008	0.012
K	3.18	4.06	0.125	0.160
L	7.37	7.87	0.290	0.310
M	15°	15°	15°	15°
N	0.51	1.02	0.020	0.040

NOTES:
1. LEADS WITHIN 0.13 mm (0.005) RADIUS OF TRUE POSITION AT SEATING PLANE AT MAXIMUM MATERIAL CONDITION
2. DIMENSION "L" TO CENTER OF LEADS WHEN FORMED PARALLEL.

Table C-3. Electrical Characteristics.
(VCC = +15 V, VEE = -15 V, TA = 25°C unless otherwise noted)

Characteristic	Symbol	MC1741			MC1741C			Unit
		Min	Typ	Max	Min	Typ	Max	
Input Offset Voltage (RS ≤ 10 k)	V_{IO}	–	1.0	5.0	–	2.0	6.0	mV
Input Offset Current	I_{IO}	–	20	200	–	20	200	nA
Input Bias Current	I_{IB}	–	80	500	–	80	500	nA
Input Resistance	r_i	0.3	2.0	–	0.3	2.0	–	MΩ
Input Capacitance	C_i	–	1.4	–	–	1.4	–	pF
Offset Voltage Adjustment Range	V_{IOR}	–	±15	–	–	±15	–	mV
Common Mode Input Voltage Range	V_{ICR}	±12	±13	–	±12	±13	–	V
Large Signal Voltage Gain ($V_O = ±10$ V, $R_L \geq 2.0$ k)	A_v	50	200	–	20	200	–	V/mV
Output Resistance	r_o	–	75	–	–	75	–	Ω
Common Mode Rejection Ratio (RS ≤ 10 k)	CMRR	70	90	–	70	90	–	dB
Supply Voltage Rejection Ratio (RS ≤ 10 k)	PSRR	–	30	150	–	30	150	µV/V
Output Voltage Swing ($R_L \geq 10$ k) ($R_L \geq 2$ k)	V_O	±12 ±10	±14 ±13	– –	±12 ±10	±14 ±13	– –	V
Output Short-Circuit Current	I_{os}	–	20	–	–	20	–	mA
Supply Current	I_D	–	1.7	2.8	–	1.7	2.8	mA
Power Consumption	P_C	–	50	85	–	50	85	mW
Transient Response (Unity Gain – Non-Inverting)								
(V_I = 20 mV, $R_L > 2$ k, $C_L < 100$ pF) Rise Time	t_{TLH}	–	0.3	–	–	0.3	–	µs
(V_I = 20 mV, $R_L > 2$ k, $C_L < 100$ pF) Overshoot	os	–	15	–	–	15	–	%
(V_I = 10 V, $R_L > 2$ k, $C_L < 100$ pF) Slew Rate	SR	–	0.5	–	–	0.5	–	V/µs

Table C-4. Electrical Characteristics.

(V_CC = +15 V, V_EE = −15 V, T_A = *T_high to T_low unless otherwise noted)

Characteristic	Symbol	MC1741 Min	MC1741 Typ	MC1741 Max	MC1741C Min	MC1741C Typ	MC1741C Max	Unit
Input Offset Voltage (R_S ≤ 10 kΩ)	V_{IO}	–	1.0	6.0	–	–	7.5	mV
Input Offset Current	I_{IO}							nA
(T_A = 125°C)		–	7.0	200	–	–	–	
(T_A = −55°C)		–	85	500	–	–	–	
(T_A = 0°C to +70°C)		–	–	–	–	–	300	
Input Bias Current	I_{IB}							nA
(T_A = 125°C)		–	30	500	–	–	–	
(T_A = −55°C)		–	300	1500	–	–	–	
(T_A = 0°C to +70°C)		–	–	–	–	–	800	
Common Mode Input Voltage Range	V_{ICR}	±12	±13	–	–	–	–	V
Common Mode Rejection Ratio (R_S ≤ 10 k)	CMRR	70	90	–	–	–	–	dB
Supply Voltage Rejection Ratio (R_S ≤ 10 k)	PSRR	–	30	150	–	–	–	µV/V
Output Voltage Swing	V_O							V
(R_L ≥ 10 k)		±12	±14	–	±10	±13	–	
(R_L ≥ 2 k)		±10	±13	–	–	–	–	
Large Signal Voltage Gain (R_L ≥ 2 k, V_out = ±10 V)	A_v	25	–	–	15	–	–	V/mV
Supply Currents	I_D							mA
(T_A = 125°C)		–	1.5	2.5	–	–	–	
(T_A = −55°C)		–	2.0	3.3	–	–	–	
Power Consumption (T_A = +125°C)	P_C	–	45	75	–	–	–	mW
(T_A = −55°C)		–	60	100	–	–	–	

*T_high = 125°C for MC1741 and 70°C for MC1741C
T_low = −55°C for MC1741 and 0°C for MC1741C

Table C-5. Noise Characteristics. (continued on page 301). (continued on page 301)
(Applies for MC1741N and MC1741NC only, $V_{CC} = +15$ V, $V_{EE} = -15$ V, $T_A = +25°C$)

| Characteristic | Symbol | MC1741N | | | MC1741NC | | | Unit |
		Min	Typ	Max	Min	Typ	Max	
Burst Noise (Popcorn Noise) (BW = 1.0 Hz to 1.0 kHz, t = 10 s, R_S = 100 k) (Input Referenced)	E_n	—	—	20	—	—	20	µ V/peak

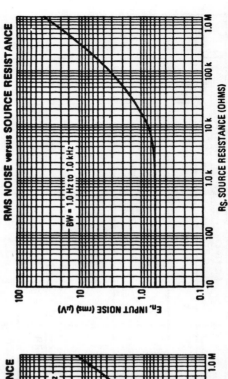

RMS NOISE versus SOURCE RESISTANCE

BW = 1.0 Hz to 1.0 kHz

E_n, INPUT NOISE (rms) (µV)

R_S, SOURCE RESISTANCE (OHMS)

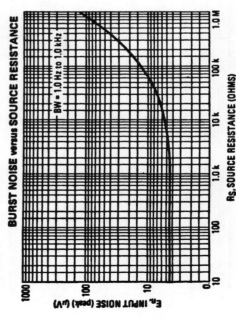

BURST NOISE versus SOURCE RESISTANCE

BW = 1.0 Hz to 1.0 kHz

E_n, INPUT NOISE (peak) (µV)

R_S, SOURCE RESISTANCE (OHMS)

Table C-5. Noise Characteristics. (continued from page 300).

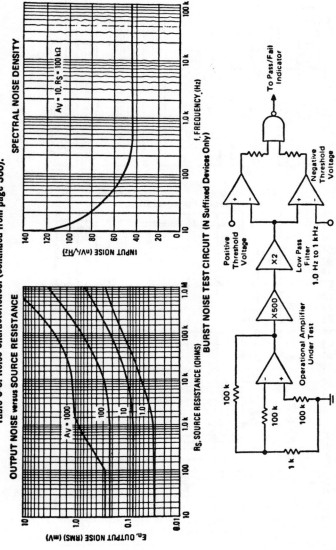

SPECTRAL NOISE DENSITY

$A_V = 10, R_S = 100 \text{ k}\Omega$

INPUT NOISE (nV/√Hz)

f, FREQUENCY (Hz)

OUTPUT NOISE versus SOURCE RESISTANCE

$A_V = 1000$
100
10
1.0

E_n, OUTPUT NOISE (RMS) (mV)

R_S, SOURCE RESISTANCE (OHMS)

BURST NOISE TEST CIRCUIT (N Suffixed Devices Only)

To Pass/Fail Indicator

Positive Threshold Voltage

Negative Threshold Voltage

Low Pass Filter 1.0 Hz to 1 kHz

X2

X500

Operational Amplifier Under Test

100 k

100 k

100 k

1 k

For applications where low noise performance is essential, selected devices denoted by an N suffix are offered. These units have been 100% tested for burst noise pulses on a special noise test system. Unlike conventional peak reading or RMS meters, this system was especially designed to provide the quick response time essential to burst (popcorn) noise testing.

The test time employed is 10 seconds and the 20 μV peak limit refers to the operational amplifier input thus eliminating errors in the closed-loop gain factor of the operational amplifier under test.

Table C-6. Typical Characteristics. (continued on page 303).

(V_{CC} = +15 Vdc, V_{EE} = –15 Vdc, T_A = +25°C unless otherwise noted)

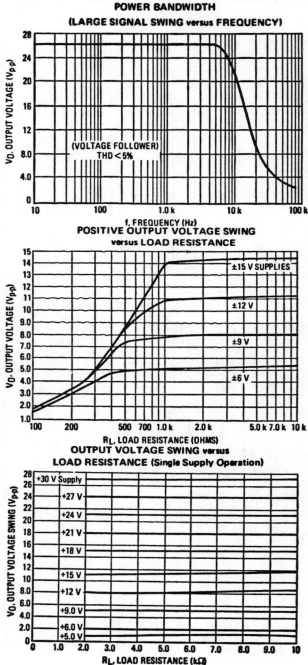

POWER BANDWIDTH

(LARGE SIGNAL SWING versus FREQUENCY)

(VOLTAGE FOLLOWER)
THD < 5%

f, FREQUENCY (Hz)

POSITIVE OUTPUT VOLTAGE SWING
versus LOAD RESISTANCE

±15 V SUPPLIES
±12 V
±9 V
±6 V

R_L, LOAD RESISTANCE (OHMS)

OUTPUT VOLTAGE SWING versus
LOAD RESISTANCE (Single Supply Operation)

+30 V Supply
+27 V
+24 V
+21 V
+18 V
+15 V
+12 V
+9.0 V
+6.0 V
+5.0 V

R_L, LOAD RESISTANCE (kΩ)

Table C-6. Typical Characteristics. (continued from page 302).

OPEN LOOP FREQUENCY RESPONSE

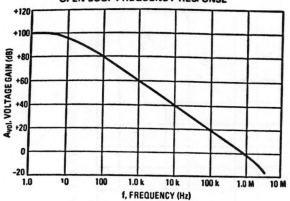

NEGATIVE OUTPUT VOLTAGE SWING
versus LOAD RESISTANCE

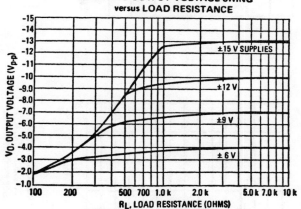

SINGLE SUPPLY INVERTING AMPLIFIER

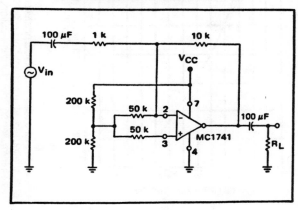

Table C-6. Typical Characteristics. (continued from page 303).

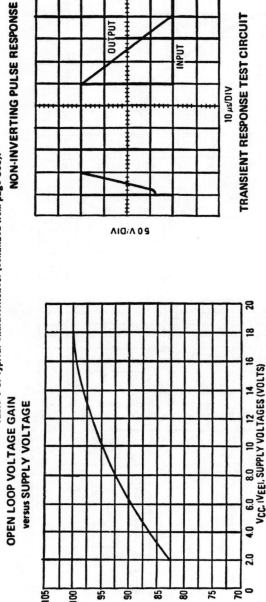

NON-INVERTING PULSE RESPONSE

50 V/DIV

10 μs/DIV

TRANSIENT RESPONSE TEST CIRCUIT

OPEN LOOP VOLTAGE GAIN versus SUPPLY VOLTAGE

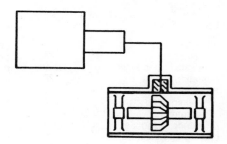

Additional Reading

LAPLACE TRANSFORM *by Holl, Maple, and Vinograde; Appleton-Century Mathematics Series.*

INTRODUCTION TO CONTROL SYSTEM TECHNOLOGY *by Robert Bateson; Merrill.*

A SIMPLIFIED TECHNIQUE OF CONTROL SYSTEMS *by Tucker, and Wills; Honeywell Inc.*

ANALYSIS AND CONTROL OF LINEAR SYSTEMS *by Y. H. Ku; International Textbook Co.*

DC MOTORS, SPEED CONTROLS, SERVO SYSTEMS *by Electro-Craft Corp.; Electro-Craft Corp.*

INTRODUCTION TO AUTOMATIC CONTROL SYSTEMS *by Robert Clark; Wiley.*

FEEDBACK AND CONTROL SYSTEMS *Schaum's Outline Series; McGraw Hill.*

AUTOMATIC CONTROL SYSTEMS *by Kuo; Prentice Hall.*

ELECTRONIC SYSTEMS THEORY AND APPLICATION *by Henry Zanger; Prentice Hall.*

THE ACTIVE FILTER HANDBOOK *by Frank P. Tedeschi; TAB Books Inc.*

PHASE LOCKED LOOP DESIGN FUNDAMENTALS (Motorola Application Note AN-535) *by Garth Nash; Motorola Semiconductor Products Inc.*

A NEW GENERATION OF INTEGRATED AVIONIC SYNTHESIZERS (Motorola Application Note AN-553) *by Richard Brubaker and Garth Nash; Motorola Semiconductor Products Inc.*

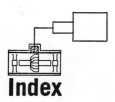

Index